Traditional Ecological Knowledge of Mountain People

Traditional Ecological Knowledge of Mountain People

Foundation for Sustainable Development in the Hindu Kush-Himalayan Region

Editors
Neelendra K. Joshi
Vir Singh

2010
DAYA PUBLISHING HOUSE
Delhi - 110 035

ISBN 9789351242161

Published by : **Daya Publishing House**
A Division of
Astral International Pvt. Ltd.
– ISO 9001:2008 Certified Company
4760-61/23, Ansari Road, Darya Ganj,
New Delhi-110 002
Ph. 011-43549197, 23278134
E-mail: info@astralint.com
Website: www.astralint.com

Laser Typesetting : **Classic Computer Services**
Delhi - 110 035

Printed at : **Chawla Offset Printers**
Delhi - 110 052

PRINTED IN INDIA

UN Mountain Agenda

"As a major ecosystem representing the complex and interrelated ecology of our planet, mountain environments are essential to the survival of global ecosystem....... The proper management of mountain resources and socioeconomic development of the people deserve immediate attention..."

☆ Agenda 21, Chapter 13

To

The mountain farming communities

whose traditional wisdom, innovativeness, holistic world-view and vision enlighten the path to heal the Planet Earth

Editorial

Knowledge is power. But mere knowledge is not power; it is only possibility. Action is power; and its highest manifestation is when it is directed by knowledge.

☆ Francis Bacon

Mountains are the islands in the sky. Their ecological contributions to the whole biosphere are life-sustaining and life-enhancing. Had there been no mountains on Earth, there would have been no river systems; and story of life, if it existed at all, would have been quite prosaic. Mountains have imparted new meaning to life; have contributed to the aesthetics of our planet, to the beauty of our cosmos. The Hindu Kush-Himalayan (HKH) Mountains constitute one of the most robust ecosystems of the planet Earth. Many of the environmental crises the planet earth is in the grip of are attributable to the very state of 'ecological coup' the HKH Region appears to be at the verge of. The HKH Region performs crucial ecosystem functions which are phenomenal for the whole globe. Any deterioration leading to retrogressive or disclimax state of the environment of the highest, the most fragile and the most sensitive mountain ecosystems of the Earth is naturally bound to phenomenally affect the global environment and, eventually, the biosphere.

A habitat bears its uniqueness owing to natural processes. When it is occupied by a human community, its uniqueness is also attributable to the human dimensions. Human species–the *Homo sapiens*–owing to its highly evolved anatomical, morphological and intellectual capabilities has brought about phenomenal changes on the planet Earth. The changes are conspicuously becoming catastrophic not only to the whole life but to the human race as well. Ushering in a state of catastrophic trend, however, is a recent phenomenon. It is largely attributable to the processes triggered by the conventional development paradigm in which nature and its functional units–the ecosystems–are regarded merely a commodity, means of economic growth and worth to be trampled upon. The current development paradigm, in essence, is life-diminishing and life-annihilating. The much talked about global warming and ominous climate changes are an evidence of the same.

Rapidly on-going liberation, privatisation and globalisation (LPG) processes have structured a new development paradigm. Regional specificities and specific ecological niches are being increasingly exploited and nature's natural elements are being rampantly converted into commodities. Everything that comes or ought to come in our way is price-tagged. Marginal communities such as those inhabiting HKH Mountains, under such circumstances, are bound to be further marginalised. The traditional knowledge systems they have evolved over millennia will prove to be a straw in the wind.

The traditional knowledge systems nevertheless have been potent tools of the marginal mountain communities. These would help local people, as they have been doing for centuries, to heal and regenerate their natural resource bases, revamp their livelihoods built around specific natural resources and respond to the LPG processes.

There is no alternative to the traditional ecological knowledge and systems based on that. Dr. NS Jodha, one of the most celebrated mountain scholars currently with the International Centre for Integrated Mountain Development (ICIMOD) based in Kathmandu, Nepal, in one of his works in the beginning of 1990s, revealed the fact that negative indicators or unsustainability indicators in mountain agriculture intensified in proportion to the institutional development intervention. This further unfolds the significance of the traditional knowledge of mountain farmers, which is evidently much better and promising than the one pushed by public sector institutions. Haplessly, the traditional knowledge of the mountain communities has been grossly neglected. It has rather been labelled as a symbol of backwardness, unscientific and a remnant of the past. Development intervention initiated by external institutions has contributed to the irreversible erosion of community-based institutions and their knowledge systems.

Despite immeasurable loss to the traditional ecological knowledge, the systems built on the foundation of this knowledge excel over those being structured around LPG processes. The traditional knowledge is rooted into ecosystem realities, is enveloped in an eco-philosophy, holds reverential attitude towards nature and nature's components, and embraces a world view of its own. Evolved by local people over millennia through trial and error, the traditional knowledge is futuristic. It is dynamic, not static. It is compassionate and caring. It is transcendental and evolves continuously and unceasingly in tune with time and space.

This volume is an effort to look into the pristine and also somewhat forgotten traditional knowledge systems in the Hindu Kush-Himalayan Region. Should the various articles collated in the volume contribute to check the erosion of people's traditional ecological knowledge and further help restore and promote the systems based on this knowledge, the major objective of this work would be fulfilled. We would very much like to appreciate the contributors of the articles who have tirelessly worked with mountain people in remote areas and have tapped very crucial information for the contemporary world. Traditional knowledge should be regarded as a heritage of a community. Its protection, recognition and promotion should be an inherent element of a policy and development intervention in an ecozone.

Hats off to the marginal HKH communities who are exceptionally rich in their traditional systems and ecological wisdom and who have kept their traditions alive to date! Deriving inspiration from the past helps us ushering in a safe and sustainable future, a future that would blossom with vibrant humanity in harmony with biosphere. We hope the book we have edited would be helpful for the mountain scholars, policy makers, discipline professionals and mountain communities to delve into ecosystem realities and work accordingly. We owe moral responsibility to cherish the values our past generations have conferred on us. It is also our pleasant duty that we enrich these values with inter-generational experiences and with new tastes of innovativeness and pass them on to the next generations.

Neelendra K. Joshi

Vir Singh

Contents

List of the Contributors

Anoop Singh
Environmental Research Institute (ERI), Lee Road, University College, Cork, Ireland

Babita Bohra
International Centre for Integrated Mountain Development, Khumaltar, GPO Box 3226, Kathmandu, Nepal

Chitra Pathak
Krishi Vigyan Kendra (KVK), Jeolikote, Nainital, Uttarakhnad, India

Deepak Pant
Environmental and Process Technology, VITO - Flemish Institute for Technological Research, Boeretang, 200 2400, Mol, Belgium

F.E. Pradhan
Institute of Forestry and Environmental Sciences, University of Chittagong, Chittagong 4331, Bangladesh

M.A. Kabir
Institute of Forestry and Environmental Sciences, University of Chittagong, Chittagong 4331, Bangladesh

M. Inoue
Department of Global Agricultural Science, Graduate School of Agricultural and Life Sciences, The University of Tokyo, 1-1-1 Yayoi, Bunkyo-ku, Tokyo, 113-8657, Japan

Manish Kumar
Intra-health, A2/35, II Floor, New Delhi – 29, India

Manisha Gahlot
Department of Clothing and Textiles, College of Home Science, G.B. Pant University of Agriculture and Technology, Pantnagar-263145, Uttarakhand, India

Mrinalini Shah
Amrpali Institute of Management, Lamachaud, Haldwani, Uttarakhand, India

Nani Ram Subedi
International Centre for Integrated Mountain Development (ICIMOD), PO Box 3226, Kathmandu, Nepal

Nargis Fatima
Department of Clothing and Textiles, College of Home Science, G.B. Pant University of Agriculture and Technology, Pantnagar, Uttarakhand, India

Neelendra K. Joshi
Department of Entomology, 501 ASI Building, The Pennsylvania State University, University Park, 16802, PA, USA

Neema Papnai
Department of Clothing and Textiles, College of Home Science, G.B. Pant University of Agriculture and Technology, Pantnagar-263145, Uttarakhand, India

O.P. Mishra
Department of Extension Education, Institute of Agricultural Sciences, Banaras Hindu University, Varanasi - 221 005 (U.P.) India

Otto Moog
Department of Hydrobiology, BOKU - University of Natural Resources and Applied Life Sciences, Max Emanuel-Strasse 17, A-1180 Vienna, Austria

Pinki Bhandari
National Centre for Integrated Pest Management, LBS Building, PUSA, New Delhi -100 012, India

Piyush Malaviya
Department of Environmental Sciences, University of Jammu, Jammu- 180 006, India

R.D. Gaur
Department of Botany, HNB Garhwal University, Srinagar-246174. Uttarakhand, India

Saurabh
College of Home Science, G.B. Pant University of Agriculture and Technology, Pantnagar-263145, Uttarakhand, India

Shachi Shah
G.B.P.U.A.T., Hill Campus, Ranichauri, Tehri Garhwal, 249199, Uttarakhand, India

Silvi Siddhu
Department of Food Science, Faculty of Agriculture and Food Sciences, University of Manitoba – Winnipeg, R3T 2N2, MB, Canada

Subodh Sharma
Aquatic Ecology Centre, Kathmandu, University, P.O. Box: 6250 Kathmandu, Nepal

Tapan K. Nath
Department of Forest Science, Graduate School of Agricultural and Life Sciences, The University of Tokyo, 1-1-1 Yayoi, Bunkyo-ku, Tokyo, 113-8657, Japan

Vir Singh
Department of Environmental Science, College of Basic Sciences and Humanities, GB Pant University of Agriculture and Technology, Pantnagar – 263145, Uttarakhand, India

Chapter 1

Traditional Knowledge Systems of Mountain Communities: Issues in the Hindu Kush-Himalayan Region

☆ *Nani Ram Subeti*

Background

Traditional knowledge of mountain people refers to innovations and practices of local communities living in different corners of the world. Human wisdom is efficient to cope with survival strategies based on physical and mental contexts of their living. People gain experiences from daily practice of their survival strategies over the centuries. These experiences have automatically been converted into a knowledge system later on and adapted in the local culture. The traditional knowledge is transmitted orally from generation to generation by the means of stories, songs, folklore, proverbs, cultural values, beliefs, rituals, community laws, local language, and agricultural practices. Evolution process of the traditional knowledge starts from practical experiences where as modern knowledge starts from creation of a theory. Therefore, traditional knowledge has been developed from practice but not from a theory (Subedi, 2001).

Mountain communities are very rich in traditional knowledge systems. While different systems of Hindu Kush-Himalayan communities are analysed, there is a greater degree of diversity in knowledge systems. Modern literature cites that Himalayan communities are isolated, unaware, and backward. If modern knowledge system is considered as a single parameter of awareness mapping, this citation may be true. If traditional knowledge system is valued as a foundation of modern knowledge system, Himalayan communities are not completely isolated in terms of awareness, education and communication. These people are having their own way of communication which is, in some cases, more effective than modern means of communication. Therefore, this article attempts to present some of good examples of traditional knowledge systems of Himalayan region which are relevant to awareness building and educating people.

Importance of Traditional Knowledge

Mountain communities at present are facing a transitional stage in knowledge system. They are heavily influenced by modern innovations and technologies brought by the concept of globalization. The globalised knowledge based system is not only bringing good innovations to the mountains but also bringing many challenges together. On the other hand, mountain communities are not able to give up the traditional knowledge system completely because these are deeply rooted and practically proved ones. Many mountain communities, particularly in the Hindu Kush-Himalayas, are following both modern as well as traditional knowledge system in their education and awareness.

At preset, not only traditional knowledge carriers but also modern knowledge brokers are realizing the value of traditional knowledge. Some of collective arguments from both sides claim that traditional knowledge is valuable not only to those who depend on it in their daily lives, but to modern industry and agriculture as well. For example, many plant based medicines and cosmetics, are derived from traditional knowledge system. There are many other products in the market today which are produced adapting some forms of traditional knowledge system.

Another view point today is that traditional knowledge can make a significant contribution to sustainable development. Particularly in mountain areas, local communities are located in areas where the vast majority of plant genetic resources are available. Mountain communities have protected and used biological diversity in a sustainable way for thousands of years. Later on, some of mountain communities have also started cultivation of such important resourses to meet their survival needs. In overall, the contribution of mountain communities to the conservation and sustainable use of biological resources goes far beyond their role as natural resource managers. Knowledge and skills of mountain people, which are mostly in unwritten form, provide valuable information to the global community and government policy makers. For example, mountain people have been following a need based communication system for a long time. They may not communicate about their activities to their leaders and subordinates for a long time if it is not required. A common understanding is that communication is not taking place because there is no need. It means that the activities are running smoothly in a way and speed as it was in the time of earlier communication. This is also another way of communication. One time effective and two ways communication is better than frequent but ineffective and one way communication. Mountain communities follow this approach very much. They do not believe on one way communication because they are traditionally practicing two ways communication in most of the cases.

Use of Visual Sense

A growing respect for traditional knowledge has led modern science to adapt its procedures for assessing the impact of development interventions in mountain communities. These communities have traditionally been using a visual sense in monitoring of ecosystems, species, particular resources and sustainable management of biological diversity. Governments are also seeking to involve indigenous and local communities more actively and to apply their knowledge and technologies in the conservation and sustainable use of such resources, but the gap is that governments want to bring theories in practice and mountain communities want to theorise their practical knowledge.

Visual means of communication are becoming more effective in the modern development age. If traditional practices of mountain communities are analysed, we find that mountain communities used to have more visual means of communications than at present. They did not have audio-visual means like television and radio today but they were having more visual sense than today. The mountain people were fond of listening music, folk songs, gathering with colourful dresses, particular dress to give particular message, etc. Despite many modern means of communication even in the mountain areas today, the mountain communities are still using these means in many places. If we look at the evolution of communication technology in the world as a reality, the visual sense of individuals helped to develop visual means later on. From this perspective the mountain communities are very rich.

Environmental Concerns

The environmental justice movement arose in a particular context in the USA and has by and large been associated with and understood in the context of race, where Afro-American communities are in a minority, although South Asia has witnessed several mass-based environmental struggles over the last several decades. In the 1990s, however, some researchers began to point to the convergence between these movements and the struggle for environmental justice in the West. However, despite the convergences, it is important to note that the struggle in the South is that of poor mountain people who constitute the vast majority of the population. However, there are significant overlaps between this vast majority and tribal, indigenous and *dalit* communities who have been marginalised from mainstream development historically. In that sense they constitute 'rural population'. When we refer to the rural population in the Himalayan context, it is not with a purely numerical reference. The struggles of these groups acquire even greater meaning and urgency today in the context of globalisation, wherein we are witnessing newer forms of resource extraction and trade-offs. Redefining environmental justice in the Himalayan context and accordingly mapping rural population is extremely topical today.

Poverty Concerns

We understand that mountain communities, as those marginalized in some way or the other, thereby including indigenous, tribal and certain ethnic groups are living on the high hills. It is widely accepted that mountain people are cited as indigenous, tribal and *dalit* and belong to minority groups. It is also true that most of these groups are socially, economically and politically marginalized and the livelihoods of these people directly depend on natural resources surrounding them. Poverty concerns in Himalayas include critical issues such as basic rights to food, potable water, material and survival. These issues are often linked to the loss of collective property rights and access to land, forests, water bodies and pastures. Projects and enterprises in the name of 'development' often have an adverse impact on the immediate environment and livelihoods of marginalized communities, whilst the benefits go to others who are living far away from the mountains.

It has been observed that mountain development programmes are not able to produce expected impacts on the lives and livelihoods of mountain communities. In many cases, mountain communities have lost their existing rights to their means of livelihood because of externally influenced development interventions. Particularly in the Hindu Kush-Himalayan region conflicting demands of natural resources have been increasing. The conflicts started from the demand of natural resources have been spreading to other aspects of life and finally getting a negative connotation in political process. However, the present realities of conflicts of the mountains have been further marginalizing mountain communities. Therefore, the people who are getting marginalized need to be identified and appropriate strategies need to be evolved in order to help them advocate for their rights for more equitable access to the resources on which their livelihoods have traditionally been depending. Today there is a critical need to empower these people so that they could find alternatives for secured and dignified livelihoods in a sustainable manner.

Dilemmas of Networking

The basic purpose of a network is to overcome from isolation in terms of successes, failure, and visibility. Network provides a platform for free flow of experiences and ideas across individuals and groups. The idea is either influenced by traditional or by modern knowledge system. Solidarity across members, sharing of good ideas, and support during crisis are important factors particularly in the mountain communities. Solidarity and mutual accountability could be either in the sense of literature or emotional. In addition, mobilizing energy and resources among members is an important area where a network can contribute significantly. The energy includes new ideas, enterprises and designs from advocacy perspectives to get the expected changes done in favour of the poor and oppressed. However, the following can be seen as dilemmas of a network in the mountain context.

Participation and Responsibility

Members of a network are always interested to participate in networks for getting news, opinion or experiences from others but they generally hesitate to take responsibilities on behalf of their communities. There are several networks representing mountain areas and traditional practices. Ideally, these networks are supposed to advocate for establishing values of traditional knowledge systems in the modern development discourses. However, the outcomes of these networks observed as of now are not much effective. Therefore, the time has come that the networks representing mountain communities should take care of this dilemma.

Coordination and Control

By definition, members of the networks remain autonomous in their activities. They are mainly responsible to ensure proper sharing and communication across network members and all those who are directly or indirectly associated with it. Inaccessibility in the mountain areas is a physical reality. There is no excuse about it and it is beyond the control of present development stakeholders at different levels. We cannot change mountains into a plain land where transportation and other facilities could be available easily. When a feeling of collectiveness comes in discussion, idea of networking is conceived immediately. However, in reality, it is often problematic because a feudal attitude is strongly rooted particularly in traditional societies. Therefore, the balance between controlling of network members and coordination among relevant stakeholders get disturbed in operational level.

Linkage between Individual and the Institution

There are two issues related with this dilemma. The first one is that the person as an individual member versus the institutional membership. Secondly, the issue is that linking a person (Chief Executive) in an organizational membership versus involving the whole institution as member organization. Networks have enormous potential to enlist individuals as members based on their interest, commitment and resources on the issues being addressed by the network. But at the same time, resources for continuing an ongoing campaign also require institutional support and therefore, both individuals and institutions are equally important.

Information Sharing and Action

Information is to be shared in order to promote further actions. Networks share information with their members with an expectation that members will act upon that information, but generally it is not taking place in the mountains. Members expect from networks that all information must be shared with them. But the general trend is that hardly a few members take action upon the information received. Similarly network secretariat receives a lot of information from members without having any idea of what to do with such information. After a certain period of time, if members find that the secretariat is not using their information for furthering their objectives, they would stop sending information to secretariat and vice-versa.

Process and Structure

Structure should be for facilitating the networking process and therefore keeping a balance between process and the structure is a challenge before most of the networks in the mountains. Networking is a process as well as an activity for achieving a common goal of mountain people. Therefore, one of the arguments about networking is that focus of the members should always be on the process. It is a fact that institutionalisation of networks is needed for sustaining that process, but that institution should not become heavier than the process. The past experiences of the mountain indicate that mountain communities are still struggling with these multi views and not getting a clear way out for effective advocacy of their knowledge systems.

Innovation and Continuation

With changing socio-economic and political scenario in all countries, new challenges are bound to happen and if networks for promoting traditional knowledge want to remain relevant for their partners and members, the networks should take up new emerging issues and challenges and work on it. Similarly, old issues, which has become important issue for the sector and got integrated into the functioning of large number of member, should be dropped from the network agenda. Dilemma arising with the staff of the network secretariat is that whether they have to develop capacity for the innovation or remaining with continuation of conventional ones.

Leadership Style

Generally it is found in the mountain areas that some network leaders are static because of their personality, acceptability, and linkages or may be that there is no space for others. This has been an experience that, unless a person dies or vanishes from the scenario, the leadership change does not take place. This results in members loosing interest because they find that they will never get an opportunity to lead the network. In the rotating leadership form of network, the leadership keeps changing which bring freshness, innovation and dynamism in the networking process. In a healthy network, every member should feel that some day they would also get an opportunity to provide leadership. Frequent changes in leadership also create problems and some times continuity gets disrupted.

Issues for Advocacy

It is widely accepted that mountain communities are very rich in traditional knowledge system. The basic question here is how mountain societies are being governed by different governing systems in the Himalayan countries. Governance, in other word, is fairness, social justice, and following transparent procedures in public affairs in the interest of the people living in certain territory. However in the case of mountain people of the Hindu Kush-Himalayan countries, the situation is reverse one. On the one hand, majority of the mountain people are out of governing processes and a few relatively well off people are–either politicians or professionals–imposing their interest to the poor in the name of development and welfare. On the other hand, the governments have only a few scope of governing in the interest of people due to increasing dependency on foreign aids, trade, technology and market (Maskay, 2000). In this situation, it is very difficult to short out the options for regaining the momentum. Nevertheless, the following options are presented here for sharing purpose.

Stand Points for Good Governance

In an ideal situation, good governance should start from the top where there are politicians in democratic government system. As representatives of the mountain communities, they are supposed to be with fair attitude towards their constituencies. In due course, people expect good policy formulation and rule of law from them.

This attitude and fairness come down to the people through bureaucrats, donors, and civil society organizations. However, during decades long experience of those organisations working with mountain people, this modality could not work properly.

In this context, it will be relevant to cite some examples from Nepal. At this moment of the history, Nepal has come at the crossroads. There is a burning need for the leaders to show that they are capable of working for the public interest. While analyzing the situation from this perspective, the Nepalese are increasingly alarmed by the fragility of the governing processes. Politics is increasingly loosing its sense of direction, control, regulation, representation and its essential roles to serve the public interest particularly in the rural areas. National integrity is slipping down to a misery. Finally, national leadership is steadily loosing its wisdom to define its vision, and understand the reality (Dahal 2000).

In this context, there is only a way that people can start fair games and give pressure to the government through civil society organizations, donors and their development programmes. However, one fact is that only the top leadership is not only responsible for not so good governance but also all other sectors are responsible. Even common people are much cultured with bad governance. Therefore, it is urgently necessary to make all stakeholders clear on the standpoints of good governance.

Monitoring of Civil Society Organization

A new trend of development interventions in some of the Hindu Kush Himalayas, particularly in Nepal, India, Pakistan and Bangladesh, is going towards NGO sector. Roughly, half of the foreign investment, particularly grant money, is channeled through NGO sector. At present, this practice is in increasing trend as a relatively less corrupted channel of investment. One of the reasons of this trend could be due to the massive corruption in government sector. Nevertheless, this is a good inclination from the perspective of civil society strengthening. At present, more nationalization of development undertakings is emerging than before. This practice has also supported to create local employment for middle level professionals. Handling development responsibilities at local level definitely support human resource development in the mountain communities for future. There is no question that matured and responsive civil society organisations can play a significant role in order to promote good governance.

However, there is a threat of mismanagement of resources by these newly coming partner organizations. If corruption is also localized, it will be even more dreadful cataclysm than at present. This is likely to happen because government mechanisms of the Himalayan countries, which are supposed to monitor national and local NGOs, are not seen capable enough to perform this job at present. One door policy of foreign funding has become a matter of debate in many countries but it has not materialized yet effectively so far. The main issue is that the governments are not looking serious enough about the fact that around half of foreign funding is channeling through NGO sector. If half of the funding could not be monitored properly, it may create another height of manipulation.

Equity Considerations for *Dalits*

The issues related to *dalits* are unique in nature throughout the Hindu Kush-Himalayas. There is a very strong caste system in these countries. Touch-ability and un-touch-ability confines of *dalits* issues are really distressing to the rural life of these countries. One school of thought argues that economy is the major phenomenon that determines the other cultural norms. If *dalits* improve their life standard, these painful social norms normally get slow down. However, another school of thought often carried out by *dalits* themselves does not agree the earlier thought.

While analyzing both the arguments, there are some rationales with both. The main problem is social awareness on traditional norms setting system. The systemic changes in traditional knowledge system are necessary rather than changes in outer arena of the problems. It rather demands more investment in education for deprived caste groups and reservations in public affairs for some years to come. The policy changes in these directions may be supportive to bring *dalits* in the mainstream of good governance.

Equity Considerations in Revenue Sharing

Equity considerations in terms of good governance refer mainly to the geographical realities. In fact, Himalayan communities have been living in difficult geo-positions, which have made human development very difficult and expensive. Revenue sharing among all geographical regions in an equitable manner is explicitly mentioned in the development plans of respective countries. The governments of these countries functioning under different political systems have also being pressured to follow the equitable policy considerations. However, this is still a matter of debate. Most mountain communities argue that service delivery from the government mechanism is not taking place in a level of reasonable satisfaction. The mountain people argue that regional development plans and region based administration scheme introduced by different political systems in the Himalayas are not able to capture the aspirations of mountain communities.

Fundamentally, equitable access to national resources is the rights of all dwellers of the nation regardless of geographical locations and economic status. The people living in various locations should be able to claim their rights rather than begging for equity considerations. Of course, society has been concentrating on responsibility of the people very much but not paying proper attention for their rights. There is also a hidden fair of the leadership that awakening people about their rights may reduce the power that local elite are enjoying at present. Therefore, awareness on basic rights of different strata of population is energy for promoting good governance in mountain communities.

Role of Watchdog Organisations

This is a top priority issue at present in all mountain areas. General understanding of mountain communities reflects if the government mechanism is not able to maintain a rule of law in all parts of countries, there is no question of achieving good governance. However, it is not clear to all stakeholders about the roles of watchdogs. In many cases, no watchdog organizations exist. Government agencies are also lacking information to be assured what is happening where in what quality. In this context, it is very important to establish and strengthen the watchdog type of organizations for different professional sectors. Realizing the needs, some organizations of this category are being grown and functioning in Nepal (Maskay, 2000). Actually at present, governance as such is lacking particularly in the rural areas of these countries. The first requisite is the presence of government in these areas. Then only the roles of watchdog organisations can be identified properly.

Conclusions

The major impression from this review is that traditional knowledge systems of the Himalayas are assesses of communities living in the mountains. Naturally, all of these traditional knowledge systems are not matching with the modern governing systems. However, there are ample opportunities of adaptation, positive modifications and inclusion so that there could be win-win situation between traditional and modern systems. To move forward to this situation, government agencies of Himalayan countries should be able to shift their roles from regulator to facilitator. It is pertinent that the government should regulate the belongings but the role of controlling over the resources should be devolved to the people concerned. Secondly, responsible stakeholders must be able to trust the mountain people and value their traditional knowledge as their strengths for development. In many cases particularly from mountain development perspective, this situation is not adequately taking place in the Hindu Kush-Himalayan region. This is the major area for advocacy initiatives in future.

Replacement of traditional knowledge by modern theory-based knowledge is not digestible to mountain communities. There are many good practices in their system, which the mountain communities do not like to loose. On the other hind, there are some practices in traditional systems, which the modern system cannot accept. For example, Khashi communities of Meghalaya, India have been following traditional governing system at village level. According to their system, women are not allowed to participate in executive councils. They are having various logics to justify why women are not allowed to participate. However, the modern participatory development and collective governance concepts cannot accept this tradition whatsoever the logics are in place. Therefore, there should be a balance between preservation and modification of traditional knowledge systems. One of the reasons of conflicts in the mountains is breaking up healthy balance.

Hence, the existing situation of traditional knowledge of the Hindu Kush-Himalayan countries at present demands that everyone can contribute for the best utilisation of practically proved knowledge. The level of contributions could be different depending upon roles and responsibilities that the person is holding but everyone can add certain values to create a healthy matching between traditional and modern systems.

References

Dahal, Dev Raj. "Towards Alternatives: Vision of Good Governance in Nepal" in *Governance in the Doldrums: Whose reality Governs Nepal*. Nepal: South Asia partnership-Nepal, 2000.

Maskay, Bishwa Keshar. *Development Governance: Agenda for Action*. Kathmandu: Center for Development and Governance, 2000

Pasteur, Kath. *Changing Organizations for Agricultural Extension in Bangladesh: Strategies for Chance*. Institute of Development Studies, UK, 2002.

Subedi, Nani Ram. *Decentralization and Development: A Case Study of Nepal*. Nepal. Ph. D Thesis submitted to Tribhuvan University under the faculty of Humanities and Social Science, 2001.

Web site: advocacy.org, "Decentralization Strategies for Empowerment" in Tools and Practice Paper # 8.

Web site: worldbank.org (1), "Empowerment and Poverty Reduction: A Source book" Poverty Reduction Group, World Bank, May 2002.

Web site: worlbank.org (2) "Institutional and Governance Review" in Tools and Practices Paper # 15.

Web site: worlbank.org (3) "Poor People's Organization" in Tools and Practices Paper # 13.

Web: unohchr.org: "Rights based Approach", office of the United Nations High Commission for Human Rights, Geneva Switzerland, 2002.

Web: the-south-asian.com/jan per cent 202003/corruption, "Corruption–Whose Money Is It Anyway? By Isodore Domnick Mendis, printed on 2/3/2003.

Chapter 2

Dairy Farming in a Himalayan Agro-Ecosystem Setting

Farmers' Articulation of Traditional Knowledge in Livestock-based Livelihoods in Uttarakhand, India

☆ *Vir Singh and Babita Bohra*

Introduction

An agro-ecosystem in the mountains functions as a unitary whole. Dairying, especially at a smallholder's scale, is one of the several dimensions of land-related activities. Basically, it is inseparably interwoven with the components of mountain agro-ecosystems–forestry, cropping, animal husbandry, and pastoralism. Among all the land-related activities operating in a farming system, dairying, indeed, is a "cream activity" with a lot of potential for socio-economic development of the mountain areas.

Smallholders in the mountains tend to depend on the diversification of agriculture in order to reduce risks and derive more economic benefits from the limited resource base they possess. Dairying is one of the most important dimensions of diversified agriculture. This ought to be the most attractive economic activity the smallholders in the mountains would like to concentrate upon.

Basic principle underlying the success of dairy farming in the plains is: Maximum livestock production from minimum private (cultivated) land. In the mountains, however, cultivated land is not a critical factor for the success of dairy farming. There is possibility of successful dairy farming even without private land resource base, which is thanks to common property resource (CPR) base in the mountain areas. In the specific context of the mountains, the principle of maximum livestock production from minimum CPR area will hold true promise for the successful efforts in dairy farming.

Dairy-Forestry-Farming Linkages

A mountain agro-ecosystem, in essence, is self-contained. There is no waste, for nutrients are recycled. A dairy farm has a key role in maintaining the system by transferring biomass energy from uncultivated land (forest/grassland ecosystem) to cultivated land/agroforestry system and recycling crop biomass into the latter. The forest/grassland ecosystem (major CPRs in the Uttarakhand Himalayan mountains) is a natural fertilizer factory as well as a dominant producer of energy within the system. It is this ecosystem that infuses "life" in the whole system and provides it an ecological stability. Dairy farm, by means of maintaining the cyclic flow-patterns, imparts the essential ecological integrity to the farming system.

A mountain farming system is truly a solar-powered agro-ecosystem in which kinetic energy received from the sun is stored as organic molecules by green plants, that, in turn, is used for plant growth and maintenance. In addition, the cropland in the mountain farming system constantly receives a 'subsidy' from the forest ecosystem (which is also solar-powered) via a dairy farm. The mountain farming system should, therefore, be referred to as a 'nature-subsidised solar-powered agro-ecosystem'. This system contrasts sharply to the one under Green Revolution cover in the plains, in which petroleum-based inputs (chemical fertilizers, pesticides, and fossil fuel-powered machines) are to be imported from outside the system. Dairy production in this system also ought to produce a lot of waste in that there is no proper cyclic flow-pattern within the system.

Relying completely on the renewable sources of energy–inexhaustible sunlight, forests/grassland biomass and crop biomass, animal and human energy, and microbial decomposition processes in the soil (indicative of soil fertility)–the mountain agro-ecosystems are controlled by ecological principles in which all the components of the farming system function as a unitary whole. Since a dairy farm mediates energy flow through lively organic linkages amongst the farming system components, it, obviously, plays a vital role in enhancing ecological sustainability of the whole system. Performance of mountain agriculture is, to a great extent, influenced by the performance of its dairy production system.

Almost all the inputs used in a traditional dairy are produced within the system. Bedding material, often in the form of pine needles, and a bulk of fodder comes from the forests/grassland ecosystems. A proportion of dry fodder (crop residues) and, weeds, green grass, and very little amount of food grains as concentrate feed come from the cultivated land. Bedding material, leftover fodder and the dung voided at stall, after being converted into manure, reach the cultivated land. This is the traditional way of managing soil fertility in the mountain agriculture. The indigenous methods of managing soil fertility (the function of microbial decomposition processes in soil ecosystem) work reasonably well under the ecological and socio-economic conditions prevailing in the mountain region. For an economical and sustainable agriculture, a dairy farm helps by recycling farm by-products more efficiently. In fact, a dairy animal is like a bio-machine that converts the plant biomass, through microbial degradation in its rumen, into almost manure forms within a short period, of about 72 hours. If the cellulose-rich plant by-products were transferred directly into cultivated land without passing through the animal rumen, their microbial decomposition in the soil would have taken much longer time and farmers could not wait for so long. A proportion of the cattle dung is voided into grazing land at the time of grazing and contributes to fertility of the soil. This would help the grasses and other forest floor vegetation to grow well.

Dairy farming also serves to supply draught animal power (DAP) to the cultivated land. DAP, indeed, is an input of critical importance in mountain agriculture. Cattle male calves at a dairy farm are maintained as drought animals. This linkage of a dairy farm with the cropland is vital for enhancing productivity and sustainability of mountain agriculture (Singh 1998).

A dairy farm supplies to a household one of the most needed and the most nutritive food items–the milk; and thus contributes to the family's health and happiness. The family uses the milk in several ways–directly, in tea and/or coffee, and as its by-products. Some of the products, especially ghee, are preservable at home.

While a traditional dairy farm in the mountains is almost 'close' with no linkages other than the inter-sectoral ones (it, however, may often export milk outside the system), a market-linked dairy farm is 'open' and occasionally imports concentrate feed, medicines, and genes (through bull semen) and almost usually exports the milk to the market and fetches money.

Dairy Production Systems

Management

Smallholders in the mixed crop-livestock farming systems in Uttarakhand have evolved two major systems: sedentary and migratory. Livestock under sedentary system of management are kept in a village throughout the year. Majority of the smallholder dairy farms operate in sedentary management. Under the migratory system of livestock management, livestock move to different locations and stay there for a defined period of time. This is an essential feature of transhumant pastoralism being practised by certain sections of societies in the region. During summer months, these animals are herded in the alpine zone of the high Himalayan Mountains. During winter season, these animals stay in the lower hills or adjoining plain areas.

One of the most important inputs in mountain farming is manure. This input is critical for sustainability of the farming system. Dairy animals play a crucial role in supplying manure to the smallholders' farms. In rural mountain areas, dairying, in fact, is inseparably associated with manure production. In Uttarakhand, virtually, entire manure goes to cropland to maintain fertility status of the soils. A proportion of dairy cattle dung is voided in grasslands during grazing. Unlike in the rural plain areas, dung in Uttarakhand mountains is not used as fuel energy, which favours soil fertility management in the system.

Drought animal power (DAP) is yet another input that is critical to productivity/sustainability of mountain agriculture (Singh 1998).

Bullocks are the only draught animals in Uttarakhand Hills and mountains, with no exception. A dairy production system involving cows essentially serves the purpose of DAP supplies to agriculture. Farmers rearing dairy cattle use the bullocks at their own farms and supplement family incomes through hiring-out plough and sale of bullocks. This system may further be referred to as dairy-manure-draught cattle production system.

Relatively higher proportion of farms following the pure buffalo-based production system indicates a preference for market-oriented dairy management. A peri-urban dairy farm and a village dairy cooperative (VDC) are mainly semi-intensive and market-oriented.

One interesting and noteworthy thing about the dairy farming in Uttarakhand is that, unlike many mountain areas in the Hindu Kush-Himalayan (HKH) Region, it is not associated with meat production, a characteristic to be shared by many other mountain areas in India. This is also true of cattle-based dairy in Nepal. Dairy, involving sacred cow in India and Nepal carries a notion of sacredness. This suggests that in order to develop dairy enterprise in areas such as in Uttarakhand, their unique cultural setting must be taken into consideration.

Dairy Species

Cattle and buffaloes are the only dairy species throughout Uttarakhand. Use of goats, sheep, and yak, as dairy animals is extremely rare. Population of cattle over the last three decades has decreased or remained stagnated, while those of buffaloes has recorded a remarkable increase. Out of a total of 1100839 animals in Uttarakhand (Haridwar district excluded), cattle comprise some 52 per cent population, and buffaloes about 48 percent. Among all the ruminants (total population 4322534), cattle and buffaloes share 45 and 21 per cent population, respectively. Some 59 per cent cows and 61 per cent buffaloes are in milk in Uttarakhand (Animal Husbandry Department 1998).

Breeds

All buffaloes in the region are indigenous. This milch species is largely an admixture of improved type Murrah and local breed. There might be a mixture of other indigenous (Indian) breeds imported from the plains but now well adapted to mountain circumstances.

Presence of pure Murrah, considered to be the most promising high yielding breed, or other pure or improved breeds, is rare. Moreover, there have been no systematic institutional efforts to improve/ upgrade buffalo breeds in the area. Cattle have been the only targets of improvement.

According to the official Livestock Census of 1993, there were 115345 crossbred cattle that accounted for little less than six per cent of the total cattle population (1978331) in Uttarakhand Hills. Out of the total cow population, crossbreds accounted only about seven per cent. Amongst the crossbred cattle population, the female comprised 62 per cent (Animal Husbandry Department 1998).

Location-Specific Characteristics

Many of the characteristics relating to a dairy farm in mountain agro-ecosystems depend upon its location. Composition and size of livestock, feeding, disease management, breeding, on-farm processing, marketing of dairy products, and many other aspects relating to the success and/or failure of a dairy also depend on its location.

Landscape, *i.e.*, whether a dairy farm is situated in a valley/ flatland or on sloppy site would make some difference. Valleys generally have plenty of water resources and so there is less likelihood of any constraint imposed by lack of water, as might be a case with a dairy farm located on a mountain slope/in an upland area.

Altitude is another vital factor that often dictates livelihood activities in the mountain areas. The type of dairy farm would also be dictated, to a great extent, by the altitude it is located at. It is the altitude that determines the vegetation type (feed resources) and microclimate in mountain areas. These two factors are crucial for the performance of a dairy unit. Vegetation type in a watershed is one of the most important factors on which the livelihood system of a society is based. At high altitudes in the Middle or Lesser Himalayas, oak (*Quercus leucotrichophora*) forms the climax vegetation of the region. Oak forests enhance the carrying capacity for the livestock significantly under sedentary management in the surrounding areas. Extensive areas under grasslands allow a society to rear livestock in great number. Farming systems at very high altitudes in the Greater Himalayas are livestock-dominated. Intensity of various animal diseases/physiological disorders, and hence their overall performance would, to a great extent, vary from location to location. Most of the dairy farms at all locations follow mixed cattle-buffalo-based dairy production system (Singh 2000, 2001; Singh *et al.* 2001).

Feed and Nutrition

Unlike in the plain areas where livestock are fed on the feed raised on cultivated land, in the mountains of Uttarakhand they mainly subsist on fodder growing in common property resources (CPRs). Civil-Soyam and Panchayat land extending over an area of 7,600 and 2,450 sq. km., respectively, are the most important CPRs in Uttarakhand (excluding the district of Hardwar). The local inhabitants also enjoy the fodder- and fuel-gathering rights on the Reserve Forests under the control of the Forest Department of the Government of India spread over an area of 23,770 sq. km. The other sources of fodder are the private property resources (PPRs), mainly the cultivated land/agro-forestry systems occupying an area (actually sown area) of 7,175 sq. km. (Singh *et al.* 1995).

Conventional animal husbandry policies generally call for growing more and more fodder crops, particularly the leguminous crops and concentrates, on arable land in order to increase animal production. But, in fact, not more than 3-4 per cent of arable land in India is devoted to raising fodder crops. In mountain areas, arable land area under fodder cultivation is almost negligible. Majority of the smallholders inhabiting mountain areas cannot afford to spare any piece of their cultivated land for fodder production. Due to the availability of CPRs in mountain areas, local inhabitants do not customarily depend on cultivated fodder. Even the large landholders would find cereal and other cash crops far more remunerative than the cultivated fodder crops (Singh *et al.*1995).

Dairy production is influenced markedly by seasonality in feed quality and availability. Farmers' preference of fodder mainly depends on the availability of particular fodder species within the accessible reach, season of foliage production, quality of the fodder, and milch animal and it's physiological state. In Uttarakhand, 15 trees and shrubs are highly preferred for fodder utilization. However, there are several other 'non-conventional' fodder species scattered throughout diverse ecological niches of the mountains. Having excellent fodder value, they are also occasionally exploited by the livestock owners (Singh *et al.* 1995).

Average annual availability of dry fodder per village is presented in Table 2.1. On an average, 83 per cent of dry fodder in the form of crop residues is grown on cultivated land (PPR) and only 17 per cent comes as grass hay from CPRs. Maximum amount of dry fodder (29 per cent) is contributed by wheat straw (share of upland wheat straw alone being 23 percent). Contribution of CPRs in terms of dry fodder (grass hay) supplies is the highest (33 per cent) in a village at high altitude upland area, which is more than twice that in a village in the valley, and at mid-altitude upland areas. Average biomass flow in terms of dry fodder from the CPRs is only 17 per cent. The rest comes from the private cropland.

Millet straws, especially the barnyard millet straw, are the most preferred ones by the livestock farmers, but our new development intervention in mountain area calls for replacement of millet crops by commercial non-fodder type crops, like white-seeded soybeans. If this sort of change were realised, the overall availability of dry fodder in the area would be seriously affected. Moreover, millets constitute major bulk of people's diet and represent unique cultural heritage of hill people. It might be suggested that millet farming in mountain areas should be ameliorated rather than discouraged. This would indirectly benefit the livestock sector in the hills.

Quantity of green fodder available to village livestock throughout the year has been presented in Table 2.2. Oak (*Quercus leucotrichophora*) leaves comprise about 12 per cent of the green fodder available for livestock. The oak fodder is extracted from the natural oak forests in the CPR areas. *Bhimal* (*Grewia optiva*) and some other fodder trees (*e.g.*, *Celtis australis*, *Cedrella toona*, *Ficus* spp., *Bauhinia* spp., *Melia azedarach*, *Prunus* spp., etc.) grown in the agroforestry system (PPR) contribute about four per cent to the green feeding. *Bhimal* foliage makes fodder of very high nutritive value (Singh *et al.* 1987, 1995) and is generally monopolised for feeding the dairy animals in milk. Green grass (hand-cut) accounts for 35 per cent of the total greens, while grazed grass as high as 49 percent. Tree leaves amount to only about 16 per cent and green grasses as high as 84 per cent of the total greens. Contribution of green grasses (during the rainy season only) to green fodder supplies is, thus, enormous. Pattern of green fodder feeding in the mountains, largely, depends on the type of ecological niche, which would exhibit specific biological spectrum.

Table 2.1: Annual Availability of Dry Fodder in Villages at Different Locations in Uttarakhand

Source of Fodder	*Straw-Grain Ratio**	*Dry Fodder Production Per Village, q (Per cent of Total)*			
		Valley	*Mid-altitude Upland*	*High Altitude Upland*	*Average Per Village*
Upland Rice	1.18:1.0	148.33 (6.32)	292.90 (8.40)	96.29 (10.34)	179.17 (7.95)
Lowland Rice	1.28:1.0	402.56 (17.15)	—	—	134.19 (5.95)
Upland Wheat	1.40:1.0	340.06 (14.49)	958.05 (27.48)	238.70 (25.64)	512.27 (22.72)
Lowland Wheat	1.37:1.0	413.88 (17.63)	—	—	137.96 (6.12)
Finger Millet	1.50:1.0	248.85 (10.60)	733.95 (21.05)	109.20 (11.73)	364.00 (16.14)
Barnyard Millet	2.00:1.0	304.98 (12.99)	908.70 (26.07)	122.40 (13.15)	445.36 (19.75)
Barley	1.50:1.0	81.00 (3.45)	68.40 (1.96)	—	49.80 (2.21)
Maize	2.00:1.0	74.88 (3.19)	—	55.00 (5.91)	43.29 (1.92)
Grassland (Hay)	—	333.00 (14.19)	524.00 (15.03)	309.38 (33.23)	388.79 (17.24)
Total Dry Fodder	—	2347.54 (100.00)	3486.00 (100.00)	930.97 (100.00)	2254.84 (100.00)

Source: Based on a study of 12 villages in two milk sheds, namely, Nainital and Almora (Singh *et al.* 2001).

* Based on Singh (1995, 1998).

Table 2.2: Annual Availability of Green Fodder in Villages at Different Locations in Uttarakhand

Source of Green Fodder	*Green Fodder Supplies to Livestock Per Village, q (Per cent of Total)*			
	Valley	*Mid-altitude Upland*	*High Upland Altitude*	*Average*
Oak (*Quercus leucotrichophora*)	853.78 (3.63)	4725.00 (12.73)	4125.00 (18.93)	3234.59 (11.77)
Bhimal (*Grewia optiva*)	325.60 (1.38)	231.00 (0.62)	60.50 (0.28)	205.70 (0.75)
Other Fodder Trees*	1010.10 (4.30)	1433.25 (3.86)	187.69 (0.86)	877.01 (3.19)
Green Grass (Hand-cut)	9102.00 (38.71)	12915.00 (34.80)	6765.00 (31.04)	9594.00 (34.92)
Green Grass (Grazed)**	12221.10 (51.98)	17802.75 (47.98)	10657.63 (48.90)	13560.49 (49.36)
Total Greens	23512.58 (100.00)	37107.00 (100.00)	21795.82 (100.00)	27471.79 (100.00)

Source: Based on a study of 12 villages in two milk sheds, namely, Nainital and Almora (Singh *et al.* 2001).

*: The other fodder trees managed in the agroforestry systems often include *Celtis australis*, *Cedrella toona*, *Ficus* spp., *Bauhinia* spp., *Melia azedarach*, *Prunus* spp., and, at high altitudes, *Quercus leucotrichophora*, etc.

**: It was assumed that a grazing cattle unit grazed grass equal to the full requirement of the animal, *i.e.*, 24 kg per day, during the rainy season (July to October) when the green grass was plentiful, and only one-fourths of the animal's requirement, *i.e.* 6 kg per day, during rest of the months when there was scarcity of the greens in the grasslands.

On an average, 58 per cent of the total fodder biomass is extracted from the CPRs while 42 per cent grows on cropland (PPR). Livestock dependence on CPRs increases as we go higher in the mountains. High degree of dependence on forests and grasslands/grazing lands for animal husbandry and dairying signifies the vital role of CPR's in the mountain farming systems. CPRs' crucial contribution can further be realised from the fact that they are a rich repository of diverse green fodder species of very high nutritive value. As much as 98 per cent of the green fodder, on an average, is extracted from the CPR's (Table 2.3). Annual requirement and availability of fodder for all livestock has been summarised in Table 4. Fodder requirement

Table 2.3: Annual Availability of Fodder from CPRs and PPRs (Quintal) in the Villages of Uttarakhand

Location of the Village	*CPRs*			*PPRs*			*Total Fodder (CPRs and PPRs)*
	Dry Fodder	*Green Fodder*	*Total*	*Dry Fodder*	*Green Fodder*	*Total*	
Valley	333.00 (97.52)	13074.88 (2.48)	13407.88 (100.00)	2014.54 (16.18)	10437.70 (83.82)	12452.24 (100.00)	25860.12 (51.85/48.15)
Mid-altitude Upland	524.00 (2.27)	22527.75 (97.73)	23051.75 (100.00)	2962.00 (916.89)	14579.25 (83.11)	17541.25 (100.00)	40593.00 (56.79/43.21)
High Altitude Upland	309.38 (2.05)	14782.63 (97.95)	15092.01 (100.00)	621.59 (8.14)	7013.19 (91.86)	7634.78 (100.00)	22726.79 (66.41/33.59)
Per Village	388.79 (2.26)	16795.09 (97.74)	17183.88 (100.00)	1866.04 (14.88)	10676.71 (85.12)	12542.75 (100.00)	29726.64 (57.81/42.19)

Source: Based on a study of 12 villages in two milk sheds, namely, Nainital and Almora (Singh 2001).

Figures in parentheses are percentage of the total fodder; and, in the last column, percentage of CPR-based fodder/PPR-based fodder.

Table 2.4: Balance Sheet of Fodder for Livestock Feeding in Villages of Uttarakhand

Location of the Villages	*Total Cattle Units**	*Annual Requirement, q*		*Annual Availability, q*		*Deficit (-), or Surplus (+), q*		*Per cent Deficit (-), or Surplus (+)*	
		Green Fodder	*Dry Fodder*	*Green Fodder*	*Dry Fodder*	*Green Fodder*	*Dry Fodder*	*Green Fodder*	*Dry Fodder*
Valley	412.50	36135.00	9033.75	23512.58	2347.54	–12622.42	–6686.21	–34.93	–74.01
Mid-altitude Upland	524.75	45968.10	1149.03	37107.00	3486.00	-8861.10	–8006.03	–19.28	–69.67
High Altitude Upland	328.64	28788.86	7197.22	21795.82	930.97	–6993.04	–6266.25	–24.29	–87.06
Per Village	421.96	36963.99	9241.00	27471.79	2254.84	–9492.19	-6986.16	–26.17	-76.91

Source: Based on a study of 12 villages in two milk sheds, namely, Nainital and Almora (Singh 2001b).

* 1.0 CU= 1.2 Bullocks, 1.0 Cow, 1.5 Buffaloes, 0.5 Cow Calf, 0.75 Buffalo Calf, 0.2 Goat.

of livestock in a village is according to the standards set for hill livestock by the Animal Husbandry Department (1998). A cattle unit's (CU) requirement of green and dry fodder is 24 and 6 kg per day, respectively. The dismal state of the livestock feeding in the area is that the requirement in each village is far more than the actual availability of fodder. Average shortage of green fodder is 26 per cent and that of dry fodder as high as 77 per cent. Shortage of may reflect lower rates of fodder extraction from the CPR areas for keeping pace with ecological regeneration of the ecosystem.

From this fodder balance, it is clear that the whole of livestock population in Uttarakhand Hills is underfed. This situation, obviously, becomes the major reason for performance of livestock in the region. Overall paucity of green and dry fodder in the Uttarakhand, according to Animal Husbandry Department (1998), is 68.18 per cent each.

Livestock farmers have their own ways to respond to the scarcity of fodder. They would spare lion's share of the nutritive fodder for a dairy animal in milk stage, would not provide any fodder to the ovine species while tied at stall at night, and would spare some food grains and purchase concentrate for milch animals. In many mountain areas, they are even evolving alternate management systems, *e.g.*, stall feeding, planting of fodder trees close to the homestead, cultivating grasses on private land unsuitable for growing food crops or vegetables (Tulachan and Neupane 1999).

Fodder supplies to livestock in the region can be improved substantially by increasing the productivity of cropland, especially through agroforestry systems and by efficient management of the CPRs. An alteration in the feeding management inside the animal sheds would save the fodder from wastage. Most of the concentrate feed utilized by dairy animals is home-produced. However, the Milk Unions also supply concentrate animal feed to the members of VDCs. Barnyard millet, barley, black soybeans, wheat bran, wheat flour, etc. make the home-grown concentrate feed.

A dairy farm, on an average, provides only 100 kg concentrate feed to the dairy animals in a year, a bulk of which (77 kg) is produced at home. A dairy farm in a VDC will feed much more concentrates, both home-produced and purchased, to the dairy animals than its counter part in other villages (Table 2.5). There has been a phenomenal 300 per cent increase in the sale of concentrate feed to

Table 2.5: Concentrate Feeding to Dairy Livestock in Sample Villages

Location of the Villages	Per Village Per Year, q			Per Farm Per Year, kg			Per DCU[2], kg	
	Home-grown	Purchased	Total	Home-grown	Purchased	Total	Per Year	Per Day
Valley	74.81 (83.50)	14.78 (16.50)	89.59 (100.00)	80.88	15.98	96.86	32.28	0.09
Mid-altitude Upland	94.88 (75.21)	31.27 (24.79)	126.15 (100.00)	72.29	23.82	96.11	38.75	0.11
High Altitude	37.16 (78.93)	9.92 (21.07)	47.08 (100.00)	54.05	14.43	68.48	23.78	0.07
Near Market Area	67.96 (83.56)	13.37 (16.44)	81.33 (100.00)	76.57	15.06	91.63	30.44	0.08
VDC[1]	85.48 (67.56)	41.05 (32.44)	126.53 (100.00)	100.56	48.29	148.85	46.23	0.13
Average	72.06 (76.55)	22.08 (23.45)	94.14 (100.00)	76.87	23.52	100.39	34.30	0.09

Source: Singh (2001)

[1] Village Dairy Cooperative

[2] 1.0 DCU (Dairy Cattle Unit)= 1.0 Cow, 1.5 Buffaloes, 0.5 Cow Female Calf, 0.75 Female Buffalo Calf

Figures in parentheses are percentage of the total.

the VDCs by the Milk Unions in Uttarakhand (erstwhile UP Hills) from 1989-90 to 1998-99 (Dairy Development Department 1998). Concentrate feed is generally fed to the animals in milk, and often monopolised for the buffaloes in lactation stage.

Dairy Development Issues and Strategies

Dairy farming is one of the most promising enterprises for smallholders inhabiting a watershed in the mountains of Uttarakhand. An increasing trend in milk production, marketing, and consumption rates in Uttarakhand is evident (Singh 1999, 2000, 2001; Singh et al. 2001). The natural resource base including vast areas under CPR regime, the huge population of dairy animals and their unique and highly adapted breeds, and diverse animal production systems holds the key to dairy development in the region. Due to the valuable self-containment feature of the farming systems in the region, almost all inputs regarded indispensable for dairy development processes grow within the system itself. Small holder dairy farmers may be backwards from economic standards, but in terms of the natural resource base they have access to, they are richer than even their counterparts in the mainstream plain areas.

The main problem the dairy sector in the region faces, however, is the low productivity of dairy herd and production systems. The large dairy animal populations and natural resources serving as rich repositories of quality fodder remain under-exploited by the dairy sector. Institutional interventions focus on three aspects of dairy production: crossbreeding, health care, and fodder production. Nevertheless, the conventional approach is not perspective-based. Small holders are not participating in the process and, therefore, the institutional strategies have shown little, almost no, visible impact on the transformation of dairy production systems. The impact has been phenomenal only on the marketing sector.

The Himalayan region of Uttarakhand represents a unique example of community-based farming system. The conventional intervention can be successful only at a family level, not at community level. Since the majority of the Hill farmers is comprised of the smallholders, they cannot meet the requirements to sustain the production flows the new development models seeking panacea type solutions offer. Any programme unacceptable at community level carries the potential dangers of socio-economic inequity among the people of a community.

Dairy development with a focus on natural and livestock resource base will be the most appropriate perspective-based strategy for the small holder-based community of the region. The most severe constraint to dairy production in the region, as we have noted earlier, is inadequate feeding of dairy animals. Milk yields of both cow and buffaloes could be increased by feeding them adequate amount of green fodder obtainable from CPRs, especially the forests. Small holders get CPR-based feed free of cost, which results into reduced cost of dairy maintenance and consequent increase in the gross returns to the farm (Singh 2000, 2001).

Natural Resource Management

Uttarakhand mountains have large areas under uncultivated land covered with forests, grasslands, scrub (poor forest cover), and perpetual snows. A large area under forests and grasslands comprises CPRs. The natural resource base endowed with diversity of fodder yielding plants is the best bet for dairy development in the region. Natural resource management, in fact, is the most important issue relating to dairy development in Uttarakhand. This would not only be the key area to evolve "green" and sustainable dairy production systems, but would also infuse vitality in the whole farming system by increasing production flows from the natural resource base and strengthening organic linkages amongst different sectors (forests/grasslands, cultivated land, livestock, and households). Multiple benefits to be accrued to the local community through regenerated natural resource base would also contribute to strengthening livelihood systems in the region.

An efficient natural resource management alone could lead to an estimated two-and-a-half times increases in milk production. Some of the critical areas that ensure ecological integrity of the resource base and need to be properly addressed are briefly discussed below:

Afforestation

Large-scale degradation of forests resulting into reduced density cover, scrubs, and barren lands have been the issues of deep environmental and socio-economic concerns. Reforestation and afforestation involving indigenous, site-specific plants with community participation will help regenerating the natural resources in the region.

Protection

Natural and regenerated forest areas must be protected against physical and biological factors. This will further enhance carrying capacity of the uncultivated area.

Increasing Diversity

Monocultures in forest areas are often the most common scenario of mountains. Such areas must be transformed into heterogeneity harbouring tress, shrubs, and herbs of numerous species through intercultures followed by protection. Such a forest area provides higher quantities of fodder for animals.

Deferred-cum-Rotational Grazing

Overgrazing often leads to ecological degradation of grasslands/grazing lands and forests particularly due to trampling of new seedlings of trees by the grazing animals. The most effective way to control the effect of grazing and keep pace with the ecological regeneration of the area is, what is known as, the rotational-cum-deferred grazing. Grasslands/grazing lands should be enriched by the intercultures of palatable and nutritive legume species.

Practicing Stall Feeding

Popularising stall-feeding would be an energy-saving/labour-reducing practice and would further help regenerate forest and grassland areas.

Alternate Energy Sources

Large-scale extraction of fuel wood has been one of the most important reasons for ecological degradation in the region. Use of more energy-efficient stoves and feasible and affordable alternate sources of energy, such as liquid petroleum gas, biogas, solar energy, wind energy, electricity, etc. would be instrumental in saving and regenerating forest areas.

Efficient Resource Utilization

Resource extraction rates should not exceed the rate of ecological regeneration and carrying capacity of an area. There has been an exemplary practice of equitable distribution of forest-based resources, *e.g.*, grass hay, in Uttarakhand mountains for a long time. This practice should be kept alive and, wherever necessary, strengthened.

Improved Composting Technique

Tree leaves are used as bedding material in animal sheds in the Hills. Bedding, mixed with dung, is composted for transferring into soil. Composting technique in the mountains, nevertheless, appears to loss tremendous energy and nutrients. Improved composting techniques would help enhance soil fertility and fodder output from croplands.

Livestock Resource Base

Livestock-dependent communities inhabit mountain ecosystems. These livestock form the core of livelihood systems mountain communities have evolved over centuries. Many non-descript dairy animal breeds found in the mountains may excel the so called high-yielding crossbreds or exotics, particularly in respect of some traits, like energetic efficiency of milk production, adaptability to the specific environment, feed conversion efficiency, resistance against certain diseases, draughtability, and the likes. Scientific characterisation and economic valuation of the specific Hill breeds that so far remain non-descript is a crucial research issue. Research into their highly desirable but neglected traits also seems urgently warranted.

Yaks, unlike in some other Himalayan areas, so far remain non-utilized in Uttarakhand. This multipurpose animal is highly suitable for high Himalayan areas. Moreover, yak is a regular breeder, may live up to 40 years of age and may give birth to 20 or even more offspring (Negi 1990). Cheese of yak milk is very popular in Nepal. Possibility of its adoption by the high Himalayan communities in Uttarakhand needs to be explored.

Revival and strengthening the cost-effective and well-proven ethnoveterinary practices as part of the animal health management in the Hills will be a novel approach for an inaccessibility- and poverty-ridden, but the indigenous knowledge-rich region, like Uttarakhand. This, therefore, is an exciting issue the validity of which should be researched in the context of modern medicine.

Technological Options

Agroforestry systems are not only environment-friendly but also contribute to fodder supplies necessary for sustained dairy production. Some of the best known fodder trees, *e.g.*, *Grewia optiva*,

Celtis australis, Bauhinia spp., etc. occur mainly in the agroforestry systems. Some more suitable indigenous trees and shrubs that yield fodder of high nutritive value and digestibility also need to be identified and incorporated in the agroforestry systems in the Himalayan mountains.

Seasonality of fodder supplies coupled with acute shortage of fodder and low rates of concentrate feeding that impose very severe constraints on the smallholder dairy production systems call our attention to formulate balanced rations incorporating local feed resources and asses them on the farm. Feeding chemically treated dry fodder and supplementing urea, molasses, and mineral mixture in animal's diet and designing new feeding systems ensuring no or little wastage as well as increased utilization of available feed and application of fodder preservation methods could be of vital importance to increase dairy production in the region.

Applying breeding techniques aimed at reducing first calving age, increasing lactation length and productivity, and decreasing dry period, and taking advantage of modern veterinary advances to control some prevailing dairy animals' problems, like parasite infestation, infertility, etc. are yet another relevant areas for research intervention. Getting dairy animals rid of all sorts of external and internal parasites through effective medicines on regular basis could lead to considerable increases in milk production.

Institutional Intervention

Despite huge investment by the Public sector, a proportional growth in dairy sector in the mountains has not been realised. The main drawback is that institutional policies and programmes are not aimed at conservation/management of natural and indigenous livestock resource bases. Market-oriented production is necessary to ensure income opportunities for the small holder dairy farmers. However, market-oriented dairy without conservation of the resource base is no effort towards evolving sustainable dairy production systems. Protection of communities' ownership on CPRs and people's participation in the management (conservation and utilization) is the must to contribute to the sustainability of dairy production systems.

Dealing with the issue of characterisation of specific local breeds, especially of cattle, might be helpful in advocating an appropriate, participatory, and mountain perspective-based breeding policy ensuring conservation of local breeds of unique traits. An implementation of such a breeding policy could be instrumental in ameliorating dairy systems in the region.

An integrated animal health management involving both traditional and modern systems of treatment aimed at providing adequate health cover to all animals, addressing particularly the inherent problem of high degree of inaccessibility of mountain areas, would be a cost-effective and accessible system for small holder dairy farmers.

Quality control of dairy products in informal market and some rules and regulations to be imposed on milk suppliers is an issue of great public interest. Formation of a framework based on intensive information and suggestions from large number of consumers can be helpful in this direction.

Continuous exposure of small dairy farmers to dairy-related education and training, provision of subsidies and credit, remunerative prices of the produce, together with awareness about health and hygiene among consumers, can create a dairy revolution in the mountain region.

References

Animal Husbandry Department (AHD). 1998. *Livestock Census Report*. Lucknow: Animal Husbandry Department, Government of UP.

Dairy Development Department (DDD). 1997. *Dairy Development–Uttarakhand: Proposed Draft IX Five Year Plan (1997-2000)*. Srinagar (Garhwal): Dairy Development Plan–Uttarakhand.

Negi, G.C. 1990. *Livestock Development in Himachal Pradesh: Retrospect and Prospect*. Mountain Farming Systems discussion Paper No. 7. Kathmandu: ICIMOD.

Singh, R. 1997. *Economics of Livestock Production System in Himachal Pradesh*. Shimla: Agro-economic Research Centre, Himachal Pradesh University.

Singh, R. 1999. 'Smallholder Dairy Farming Initiatives: Success and Failure of Milk Cooperatives in the HKH'. Paper presented at the International Symposium on *Livestock in Mountain/Highland*

Production Systems: Research and Development Challenges into the Next Millennium held in Pokhara, Nepal from 7-10 December 1999.

Singh V. 1998. *Draught Animal Power in Mountain Agriculture: A Study of Perspectives and Issues in Central Himalayas, India*. Mountain Farming Systems Discussion Paper No. 98/1. Kathmandu: ICIMOD.

Singh, V. 1999. 'Trends and Patterns of Smallholder Dairy Farming in UP Hills'. Paper presented at the Workshop on *Smallholder Dairy Farming in the Mixed Crop-Livestock Farming Systems in the HKH Region*, jointly organised by ILRI and ICIMOD at Kathmandu, Nepal on 9-10 August, 1999.

Singh, V. 2000. *Smallholder Dairy Farming in Mixed Crop-Livestock Farming Systems in Uttar Pradesh Himalayas, India*. Kathmandu: ICIMOD

Singh, V. 2001. Sustainable Dairy Farming in Uttarakhand, India. Paper presented at the Meeting on *Smallholder Dairy Farming in the Hindu Kush-Himalayas*, Dhulikhel, Nepal, April 18-20, 2001.

Singh, V.; Sharma, R.J. and Kumar, A. 1995. 'Perspectives on the Utilization of Forest Fodders in the Mountains'. In Singh, V.; Sharma, R.J. and Kumar, A.(eds) *Ecological Carnage in the Himalaya*. Dehradun: IBD, 151-177.

Singh, V; Tulachan, P.M.; and Partap, T. 2001. Smallholder Dairy Farming in the Mountains: Potential for Operation Flood in the Uttarakhand Himalayas, India. Paper presented at the International Symposium on Mountain Agriculture in the Hindu Kush-Himalayan Region, Kathmandu, Nepal, May 21-24, 2001.

Tulachan, P.M. 1998. *Livestock Development in Mixed Crop Farming Systems: Lessons and Research Priorities*. Issues in Mountain Development 98/5. Kathmandu: ICIMOD.

Tulachan, P.M.; and Neupane, A. 1999. *Livestock in Mixed Farming Systems of the Hindu Kush-Himalayas: Trends and Sustainability*. Kathmandu: ICIMOD and FAO.

Chapter 3

Indigenous Knowledge for Sustainable Mountain Development: Some Examples from the Himalayan Region

☆ *Manish Kumar and Chitra Pathak*

Introduction

People living in the Himalayan region have accumulated tremendous indigenous knowledge (IK) and experience over the years. The area of knowledge ranges from agriculture, natural resource management, health, education, to livestock management. This rich repository of traditional knowledge has served the purpose of these communities and the larger human civilisation for centuries. Virtually every scientific and social discipline can find valuable insights in the centuries-old knowledge base of the indigenous peoples. Similarly, indigenous peoples are willing to discover what is useful and apply it to their own needs. This two-way flow of knowledge can only strengthen our mutual capacities to inch closer to sustainable and equitable development *(Eileen Conway, Editor-in-Chief, IDRC Reports Magazine).* Exchanges between indigenous peoples and Western–knowledge systems can be mutually beneficial.

Unfortunately with the advent of modern science and technology, indigenous knowledge finds it cumbersome to maintain the pace with and justify the parameters of scientific knowledge. As a result either it is completely taken over or sidelined by the proponents of modern scientific knowledge. However, the indigenous communities in the Hindu Kush Himalayan (HKH) Region have nurtured and sustained their knowledge for over many centuries primarily through a rich oral tradition.

Despite witnessing the industrial and information revolutions around the world, the mountain region remained deprived of technological movements affecting the masses. This does not imply that the benefits of technological innovations have not reached the people in the remote areas of the mountains at all. Rather the impact is minimal and majority of the population is yet to experience livelihood changes similar to those experienced by people residing in other parts of the country. Tough geographical terrain, lack of infrastructure and livelihood opportunities, chronic poverty, unskilled work force, and resource constraints are the primary obstacles to the development of the remote areas of the HKH region. Nevertheless, the realisation that indigenous knowledge is a key to the survival of the indigenous communities and for the ecological sustainability is becoming stronger.

Indigenous peoples are the original practitioners of sustainable development, applying time-tested practices to establish an ecological balance with their environment. Climatic changes have indeed raised new challenges to the indigenous sustainable development practices in the mountains. Then how do they survive?

People in the mountains adapt their skills and practices according to their growing and changing needs. Adaptation mechanisms are the primary agency ensuring the survival of people in the mountains. Such mechanisms are in use until they fail to work. Unmanageable and inadaptable crises situations force people to migrate to plains or urban areas in the hills.

Emphasis on mainstream communities and societies in the remote areas is contributing towards the disappearance of indigenous knowledge and communication systems. This loss is a serious concern as it is the repository of our ancient and time-tested knowledge, which shows the roadmap for living in harmony with the nature.

Any project or programme, which starts with the assessment of existing indigenous knowledge and skills, has better prospects of achieving sustainability in the long run. Since it builds on the readily available, cost effective, viable and locally relevant knowledge and practices, the sense of ownership and belongingness is greater. The participation of community members in development processes is enhanced through the elicitation of indigenous knowledge. The sense of legitimacy of their own knowledge and belief systems that arises from such introspection can contribute to the empowerment of these communities, assisting them to resist unwelcome outside interventions and play more active roles in determining their own future.

The industrial growth and IT revolution clubbed with globalization and liberalization have made the indigenous knowledge accessible to market forces. The threat of the commercial exploitation of the indigenous knowledge and the community is greater than ever before. Private companies often take advantage of the ignorance of indigenous people and use their knowledge for profit making ventures.

Local people involved in the collection and documentation of indigenous knowledge can become active partners in the subsequent research and development processes. Besides, input intensive technologies developed for lowland ecosystem are harmful to the highland ecosystem of the mountains.

Searching for indigenous knowledge necessitates that the researchers acknowledge the role that different members of a household or community play and the accompanying knowledge that is related to those roles. Women, for example, usually possess significant knowledge regarding agroforestry, medicinal herbs, animal husbandry, etc, which is usually overlooked and discounted by the researchers (www.icimod.org).

Difficult living conditions, deteriorating environment (like landslides, earthquakes, water shortage, etc.) and lack of livelihoods opportunities have resulted in an increased out-migration of people. This in return threatens the social system and cultural identity of the mountains.

In the light of the above discussion this article has tried to reinforce the vital role of women in the sustainable development of the mountains besides describing *Jhum* cultivation from the IK perspective and sharing some examples of IK and practices. It has also emphasised the need for digitisation of IK as it would protect it from unauthorised commercial exploitation, facilitate easy sharing and preserving it for the benefit of generations to come.

Maiti Forest: A Case of Mountain Women's Traditional Knowledge

Traditional knowledge is still the mainstay of the society in the state of Uttarakhand. It provides locally manageable, sustainable and cost effective survival strategies for local community, especially to the poor and marginalised. Women's knowledge has been the driving force for agriculture, animal husbandry, dairy and forestry. They possess good knowledge about the collection, storage and use of seeds as shown by a study done by Sahayog, an NGO based at Almora (Das,1995).

Women have been at the forefront particularly in forest conservation movements in Uttarakhand. The idea of *Maiti* forest came into existence as a result of people's, (particularly women) concern to conserve the forest wealth, which gave birth to the idea of *Maiti* forest, in the Chamoli District of Uttarakhand. *Maiti* means Mother's Home and hence the name, as it starts from mother's home. The young maids of village joined and continue to join hands to build the *Maiti* Forest. It starts with the raising of nurseries and formation of *Maiti* groups comprising of village girls. The involvement of young girls in this movement has attributed it another name that is *The Movement of the Daughters of Himalaya* (Kunwar, 1999). On the occasion of the marriage of any *Maiti* girl, the plantation of saplings from the nursery is the responsibility of other girls. The pits are dug near the bride's house. The plantation of the tree is an essential component of every marriage ceremony. The bride presents the sapling to the bridegroom who plants it under the chanting of sacred *mantras* by the *Brahmin*. The other *Maiti* girls also plant their saplings in the village. The responsibility to take care of the plant rests with the *Maiti* girls. The tree has a special significance for the parents of the bride since it is an embodiment of their girl. Today the region flourishes with these *Maiti* trees. The *Maiti* movement has spread to hundreds of villages bringing to the fore the same idea.

During the course of time, the movement has imbibed the existing cultural and social practices, thus becoming part of the social-cultural system. This has happened without much resource consumption. It has spread to a larger number of villages on its own. The success of this movement lies in the strong bonding between women and forest. Her relationship with the forest is very emotional and natural.

Here is another example of how cultural practices are mingled with the conservation activities to save the diversity of the Himalayas. This cultural aspect of conserving and saving the forest wealth has far reaching effect on the people.

The programme and projects in the mountain region should pay attention to the cost effective means and methods being practiced by rural women for years for sustainable development.

Shifting Cultivation: A Case of Indigenous Farming in the Eastern Himalayas (Adapted from www.icimod.org)

Shifting cultivation or *jhum*, *taungya* or *dongya* in the Himalayan Region, is practiced in all the mountain farming groups of people in the Eastern Himalayas as a traditional knowledge system.

It has been an indigenous life-supporting system modifying both agricultural practices and the mode of life of the people, in other words, these practices have not only influenced the lifestyle of mountain people but also the mountain ecosystems. Therefore it is a reflection of man-environment relationships, which evolved along with the issues of forest management and the maintenance of biodiversity resources.

Indigenous people have developed highly complex and very specific knowledge of their local vegetation. This body of knowledge, as well as land-use systems and the conservation practices on which *Jhum* is based should be used to the fullest extent to preserve the varied mountain ecosystems.

Case studies have shown that *Jhum* communities in the mountain areas have a rich and reliable traditional knowledge system for operating as well as adjusting to their substantial life support systems. For example the inter-cropping systems of *Alnus nepalensis* and Cardamom plants established on *Jhum* fields in the Sikkim Himalayas.

Jhum agriculture has led to evolution of a large number of human uses for species of plants and wildlife in both *Jhum* and fallow fields.

For example the edible tuber legume, *Moghania vestita*, or *Sophlong* as it is known locally by the *Khasi* and *Jaintia* people in North East India. The measures to conserve crop-genetic diversity will have to take into account these *Jhum* crops, varieties, and indigenous knowledge about using and preserving these biological species with *Jhum* agriculture.

Indigenous Knowledge and Sustainable Development: Some Examples

Indigenous Land-use Systems

The indigenous land-use systems in the Indian Himalayan region take into account ecological, social, and religious factors. Since the availability of cultivable land is determined by fertility status and moisture regimes, fields are dispersed at different altitudinal zones. Land consolidation has occurred in such a manner that each family gets its share of the upper, middle, and lower reaches of the slopes (www.icimod.org).

Indigenous Knowledge About Health, Nutrition and Livestock

In a country like India, it is, probably, true that a particular technique, which has been practiced frequently for centuries is taken for granted. In a sense, people become proficient in practicing those skills and techniques. So they do not see any need for such skills or knowledge to be documented. Documentation of such indigenous knowledge is worthwhile and could serve as an important repository of knowledge. Gupta (1990) also emphasized the necessity of documenting traditional agricultural practices.

In her research study on the *Participatory Message Development for Hill Women on Health and Nutrition in Kumaon region*, Pathak (2004) investigated and documented the indigenous knowledge of the respondents related to health and nutrition. Informal discussions were held with the respondents to explore the local treatments of common health problems like diarrhoea, constipation, allergies, injuries, common cold and fever etc. Some of the respondents explained the details of such treatments. In some cases, it was felt that they were not willing to share details of their traditional medicines and practices for some or the other reason. One of the respondents told the researcher that certain conditions are to be fulfilled before the details of curing snakebite can be revealed. Like both the learner and the teacher should have a positive thinking for each other and for the practice as well. It should be taught on a full moon night that falls on Monday. The respondent further added she had cured so many cases of snakebite with some herbs and *mantras* (sacred texts). She pointed out that for curing snakebite, both herbs and *mantras* are equally important. Interestingly, she also mentioned that to retain the effectiveness of such treatments, they should not be explained to everybody and confidentiality should be maintained.

The indigenous knowledge and practices gathered through such informal discussions have been described under the following headings reflecting different aspects of health:

- ✰ Digestive system
- ✰ Common diseases/allergies
- ✰ Fracture/injuries
- ✰ Localised pain
- ✰ Infants
- ✰ Livestock

Treatment for ailments of digestive system

Diarrhoea

The respondents explained three different alternatives for the treatment of diarrhoea:

Constipation

For constipation, they suggested the following treatments:

(*a*) Part of "*Harar*' (*Terminalia chebuls*), *Baher* (*Termminalia belerics*) and *Amla* (*Emblica officinalis*) is given along with salted (Black salt) water.

(*b*) Crushed root of '*silphar*' grass with boiled water.

(*c*) Crushed root of '*Gathiya*' grass with boiled water.

Stomach Pain

They reported that people use '*jwan*' (ajwain) with boiled water during stomach pain.

Mucous

For treating mucous, they mentioned two alternative treatments.

(*a*) *Bhatt* (cooked rice) is given with Ghee.

(*b*) Crushed root of '*karua*' grass is given in a ratio of 3:5 with leaves of *Bamani* (Brahmni).

Treatment for Fever and Common Cold

To deal with the problem of fever and common cold, the respondents described four different options.

(*a*) Ginger, black pepper and leaves of guava are boiled in water/milk and a concentrated solution is given to the concerned person before sleeping.

(*b*) *Tulsi* leaves, cloves, black pepper, ginger, sugar and *neem*, leaves of guava and '*gulbansta*' are boiled with water. This mixture is given to the infected person.

(*c*) Leaves of '*gulbansta*' and *tulsi,* together with honey are given.

(*d*) A teaspoon full of the juice of ginger, black pepper powder and honey are offered.

Treatment for Skin Diseases/Allergies

To control skin diseases, the respondents reported the following two methods.

(*a*) A complete bath with cow urine.

(*b*) Bath with water having *neem* leaves.

Treatments for Fracture/Injuries

In case of small injuries or fractures, they reported following curative measures.

(*a*) Paste of the stem of *Harjojan* (*Vitis repanda*) *Chaullyee* and the bark of *Chanadmooh* is used as plaster.

(*b*) Milk with turmeric and *fitkari* (alum) is given to the injured person.

(*c*) Dry cow dung (powdered form) is used to stop the bleeding.

(*d*) For snakebite they use *'nirvishe'* grass on the wound. The garudpanchi/*gaarudbooti* grass is also used sometime.

Localised Pain

The respondents were using different treatments for different types of pain:

(*a*) During headache, they use leaves of *Ratpatiya* fried in the ghee to massage the forehead.

(*b*) In case of tooth pain, seeds of *teemoor* (*Xanthoxylum alato*) are used; some time garlic is also used. In both the cases *teemoor* seeds and garlic is used directly on the infected teeth.

(*c*) Juice extracted from crushed leaves of *Hazari* (rose-marigold) is used for ear pain.

(*d*) Juice extract from onions is also used for earache.

(*e*) Juice extracted from the leaves of *"Ratanjoot"* is used in case of pain due to *pus* formation in ear.

(*f*) A massage with 'gun oil" is done in the case of backache.

Treatment for Infants

(*a*) A concerted solution consisting of the leaves of *neem, tulsi, brahmni* with sugar and water is prepared and used as *ghutti.* Some drops of this solution are given for stomach pain.

(*b*) *Harad, baher* and *chippi* are used in powdered form with water for constipation.

(*c*) For constipation, crushed roots of grass *ghatiya* are also used with breast milk.

(*d*) Massage with mustard oil and *heeng* is done around the navel for treating stomach pain.

Treatment of Livestock

(*a*) In case of a foot disease locally referred to as *'Khurri'*, Geru is used on the infected foot. In such conditions, the infected animal is kept at some distance from other healthy animals.

This research study highlighted the existing usage and importance of indigenous knowledge in the life of mountain people. A systematic and continuous effort should be made to document the prevailing local practices in various area of development before they vanish.

Digitization of Indigenous Knowledge

A horde of information and communication technology tools is available today to disseminate and document information, knowledge and experiences of people and communities. Digitization of content (e-content) is an important focus area for public as well as private sector players. Growing concern about bio-prospecting and Intellectual Property Rights issues related to biodiversity conservation make it imperative for us, to document the indigenous knowledge and practices of the people. Creating databases of the flora and fauna found in the Himalayan region would not only prevent their piracy and unsolicited commercialization but also promote a convenient and wide sharing of information, knowledge and experiences among different stakeholders. Moreover, the IK and practices would be preserved for the usage of future generations and also for the development of people across the world.

Conclusions

The indigenous knowledge system has been a key to the survival of the hill society in the HKH region. It has been a source of livelihoods opportunities and ecological sustainability. Despite the emergence of modern scientific knowledge, the importance of IK in sustainable development remains. Recognition of local people, especially women, and their knowledge is important for the sustainability of any development initiative in the HKH with women playing a vital role. A number of examples that demonstrate the relevance of IK in the area of sustainable development exist.

Indigenous knowledge is unique to a culture. It is the information base for a society. Indigenous knowledge is usually codified in the language of a particular society. It is dynamic and for many years has been transferred from one generation to the next, usually by word of mouth. IK system should be protected from being lost for a number of reasons. The documentation (both in text and digital format) of IK must receive prominence in order to prevent their exploitation by market forces. Besides, it is equally important to ensure that the present and future generations benefit from a rich repository of knowledge and experiences. Thus, use of indigenous and traditional knowledge for sustainable mountain development is not an option but is an imperative.

References

Das, Abhijit. 1995. Documenting Biodiversity in Action", SAHAYOG, Almora.

Das, Abhijit. 1995. *Strengthening Traditions: Participatory Research in Traditional Knowledge Systems*, SAHAYOG, Almora.

Gupta, A.K. 1990. Documenting indigenous farmers' practices, *ILEIA Newsletter* 6(2): 29.

Kunwar, Chhaya 1999. Role of mountain women in the management of sustainable ecosystem, paper presented at the 7th International Interdisciplinary Congress on Women–Women's World, 20–26 June 1999.

Pathak, Chitra. 2004. *Participatory Message Development for Hill Women on Health and Nutrition: A Study in Kumaon Division of Uttarakhand*, unpublished PhD thesis, Department of Agricultural Communication, G.B. Pant University of Agriculture and Technology, Pantnagar, Uttarakhand.

www.icimod.org.

Chapter 4

Areca catechu and *Piper betel* in Traditional Agroforestry of Rural Bangladesh: Farmers' Indigenous Knowledge and Divergence

✩ *Tapan K. Nath, M. Inoue, F.E. Pradhan and M.A. Kabir*

Introduction

Growing of palms in association with agricultural and other plantation crops is a widespread practice on the Indian subcontinent and elsewhere (Gujral, 1991). Benefits of growing palms in terms of overall land productivity are well known to the farmers. Mixed cropping with some agricultural and horticultural species increased economic output (Jacob and Alles, 1987). Bhosale *et al*. (1994) found that intercrops in areca nut plantations increased profitability by 37 per cent over pure areca plantations. Another study reported that cardamom cultivation with areca nuts increased farm income considerably in India (Korikanthimath and Rajendra, 1994).

In Bangladesh, supari palm (*Areca catechu*) cultivation with some agricultural crops is a very familiar practice in northern and southwestern regions. Farmers developed this traditional agroforestry system spontaneously to compensate[1] for crop losses arising from various adverse climatic effects like droughts and floods. It mostly takes the form of homestead boundary plantations, raising mixed crops on relatively elevated land, and irregular planting of trees on crop land (Ahmed, 1995). Traditional agroforestry systems generate income not only for the landowner and tenants, but generate employment and income for traders, woodworkers, and artisans. The additional employment- and income-generation functions of traditional agroforestry systems play a vital role in making them sustainable (Abedin and Quddus, 1991).

Substantial literature exists on the traditional agroforestry systems of Bangladesh. Scholars, for example Leuschner and Khaleque (1987), explore the structural and functional aspects of homestead agroforestry along with farmers' preferences and attitudes toward species selection and planting, and the role of women in collecting fuel and in planting and cultivating trees, and they concluded that there are favorable conditions for successful implementation of homestead agroforestry systems. Mustafa *et al.* (2000) studied the indigenous management of home gardens in four climatic regions of Bangladesh. They found that the existing management systems operate at an effective and simple level passed on from generation to generation. However, the most intensive (also extensive) research on tree management on the homesteads and farms of Bangladesh was conducted by Hocking and others. In a series of studies (*e.g.*, Hocking *et al.* 1996, 1997; Hocking and Islam, 1998) they explored farmers' preferences for trees, gender-based roles in tree management, crop yield (both annuals and trees), and interaction with rice fields. They found that village women play a significant role in the selection of homestead tree species and take care of trees.

Though these studies have been able to determine the different functional and management aspects of traditional agroforestry practices of the country, very little is known about the traditional supari palm-based agroforestry system practiced extensively in northern rural Bangladesh. Differences among farmers in allocation of land for agroforestry, and in the marketing of products, are also important for further development of this system, which was not considered in earlier studies. Furthermore, since this traditional agroforestry system accommodates a number of plant species, and because farmers have intimate contact with the agroforesters, it is rational to assume that they have detailed knowledge of the production system and the interactions among the agroforestry components (Teklehaimanot *et al.*, 2001).

Therefore, the present study was carried out to explore the prevailing supari palm-based agroforestry system, indigenous management techniques, and differences in allocation of land for agroforestry, existing tree density, production, marketing, and income difference among farmers. Farmers' ecological knowledge regarding the interactions among different agroforestry components was investigated. Finally the constraints and potentialities of the system were studied. The data base thus obtained from the study can be used for further development and extension of the system.

Materials and Methods

Study Area

The study was conducted in the Panchagarh district of Bangladesh, lying between 26°15'–26°38' N latitude and between 88°15'–88°45' E longitude. The north, northeast, and west sides of the district are on borders with India. From an agro-ecological point of view, the area is on the old Himalayan Piedmont plain (FAO/UNDP, 1988). The region is covered by Piedmont sands and gravels which were deposited as alluvial fans of the Mahananda and Karatoya rivers and their distributaries issuing from the Terai area at the foot of the Himalayas. Deep, rapidly permeable sandy loams and sandy clay loams are predominant. The topsoil is strongly acidic, and the subsoil is moderately acidic. Organic content is generally higher than in most floodplain soils of the country. The general soil type is non-calcareous dark grey floodplain soils, coming under the inceptisols soil order under the USDA soil classification (Hussain 1992). Physiographically the area gently slopes with an altitude of 33 to 96 m.

In comparison to other parts of Bangladesh, the area is relatively drier and forest-poor, having only 2 per cent tree cover including trees growing around homesteads (Hammermaster 1981). The mean annual rainfall is only 1285 mm (±363 mm). Mean maximum and minimum temperatures are 34.1°C and 15.6°C, respectively. Mean relative humidity of the area is 72.6 per cent (FAO, 1988).

Methods of Data Collection

The northwestern part of Bangladesh is famous for the production of *supari* (fruit of the supari palm) and betel leaf (*Piper betel*). This part consists of seven districts. Among them, Panchagarh district was selected using a simple random sampling. Panchagarh district comprises five (05) upazilas (sub-districts). From these five upazilas, Panchagarh *sadar* (central) upazila was selected purposively as most of the villagers of this upazila practice the supari-based agroforestry system. All the villages of this upazila were listed alphabetically in ascending order. From these, five (05) villages were selected using a simple random method.

The study was conducted by interviewing the members of rural households and visiting their agroforestry plots. The households having supari-based agroforestry plots in each village were categorized into five distinct classes based on the extent of their total land holdings: a] poor (up to 0.20 ha), b] marginal (0.21 to 0.50 ha), c] small (0.51 to 1.0 ha), d] medium (1.01 to 2.0 ha) and e] large (more than 2.0 ha). Field data were collected from 100 households, 20 from each village (four from each category in each village). Households were selected randomly.

For each household, a semi-structured questionnaire was used to collect data regarding its land and agroforestry, land allocation for agroforestry plots, and production and income from agroforestry. The number of supari palms on each plot was counted. Farmers were asked to quantify the yield and the amount sold in markets. Though farmers keep no written records, surprisingly they could remember the quantities sold and income derived from the marketing of products. A short informal discussion with the household members in each family was carried out to learn the indigenous ecological knowledge about their agroforestry.

In each village, a group discussion was conducted using a semi-structured questionnaire. The discussion highlighted the indigenous cultivation techniques and management of agroforestry, marketing of products, prevailing problems, and future of this agroforestry system. Traditional ecological awareness was also discussed in these groups. Group discussion helped to cross-check some of the information gathered in household interviews. All interviews were done whenever it was convenient for farmers. The interviewers were well acquainted with the methods of participatory rural appraisal.

Results

Access to Land for Differential Usages and Tree Density

Farmers in the study areas have access to several types of land. Farmers categorized their land into homestead, farmland, pond, fallow, and others. Homesteads are located on relatively elevated land. A typical homestead in Bangladesh contains one or several residential buildings surrounded by land with vegetable gardens, hedges, and/or trees and shrubs. Farmland is used for rice cultivation twice a year and for *rabi* crops in winter season. *Rabi* crops include country beans, pulses, potatoes, and other vegetables. Supari palm-based agroforestry plots in the study area were found around the homestead, in farmland, and also in some fallow land. Farmers plant supari palms at irregular intervals on the pond bank. Due to the long summer drought and the difficulty of irrigating in late winter, farmers are reluctant to cultivate relatively elevated farmland for rice, so they convert a portion of their land into *supari-bagan* (supari palm-based agroforestry).

The socio-economic conditions of the people of study areas are mostly based on agriculture. A few have businesses, mainly agro-based. As a result, farmers allocate most of their land to agricultural purposes. More than 70 per cent of total land in the study area is used for agriculture (farmland), followed by 15 per cent for homesteads. From these farmland plots and homesteads, farmers spare some land for supari palm-based agroforestry. In accordance with total land holdings, Poor farmers allocated more land to agroforestry (13.58 per cent) than marginal and small farmers. Farmers of Poor categories said they shifted from pure agriculture to supari palm-based agroforestry system because they have less farmland and because of frequent crop losses due to adverse weather, and to insects and other pests. This trend is also observed in the medium and large farm categories. They practice agroforestry on a considerable portion of their land.

There was also remarkable variations among farmers with respect to tree density. Though Poor farmers have the smallest landholdings, they have the highest number of palms per ha. Tree density in their agroforestry plots was almost two times higher than those of marginal, small, and medium landholders. Agroforestry plots of Poor families are mostly located within their homesteads and near their residences.

Structure of Supari Palm-Based Agroforestry System

Components

A number of compatible crops, both annuals and perennials, are grown in supari-based agroforestry in the study areas. The morphology and growth habits of the supari palm permit other crops to be intercropped during its different growth stages. During the initial growth phase, this palm requires partial shade. Farmers plant bananas during this period. Banana plants are grown there for three to four years. When the supari palms are seven or eight years old, the plantation is ready for planting betel leaf. The betel is a perennial vine that climbs on supari palms. Farmers plant rooted cuttings with three or four nodes at the base of each palm. As the supari palms start bearing fruit, farmers cultivate other agricultural crops on space available under the palms. The choice of crops depends on market demand as well as household needs. Our survey found that in general farmers cultivate turmeric, ginger, pineapple, sweet potatoes, arum, and other crops.

Interaction of the Components

The main expectation for an agroforestry system with both annual and perennial crops is that the overall return from a unit piece of land is increased without adversely affecting either the current or long-term productivity of main (perennial) crops (Liyanage *et al.* 1989). At the same time, the returns from the additional crops should justify the adoption of inter-cropping and should contribute to the long-term productivity of the system.

Farmers explained the interaction among different components of the system based on their indigenous knowledge, which is passed on from generation to generation. According to farmers, supari palms allow sunlight to pass through to open areas on ground because they grow straight and have very small crowns. They plant at wide intervals in supari palm plantations, generally 2.0 to 2.5 m between trees and 3.6 to 4.5 m between rows. Some of the open space is reserved for perennials like pineapple, while the remaining parts are planted with annuals like turmeric or ginger These species are very small (not more than 1.0 m in height) and grow well in partial shade, according to farmers. Farmers reported that they do not fertilize or irrigate the supari palm plantations and betel, other than applying some manure and mulch. About the only way they tend the palms is weeding. Intercrops, however, are given intensive care. They plough the land, apply water and inorganic fertilizers whenever necessary to increase yields. They believe that fertilizer and water provided to agricrops is probably also shared by betel and supari palm. In this way, the whole agroforestry system benefits. Working the soil improves its texture and structure, which in turn increases soil water retention. Farmers cultivate more than one agricrop in a year. As a result, supari palms and betel get nutrients and water almost year-round. Again, though farmers use supari palms to support betel vines, they have no idea whether the vines have any positive or negative impacts on supari palm growth.

Indigenous Management System

Farmers practice supari palm-based agroforestry both on their homesteads and elevated farmland. Homesteads are well developed and comprise many plant species along with supari palms. When agroforestry is practiced on farmland, farmers use these indigenous techniques:

Management of Supari Palm Plantation

Though the socioeconomic conditions of the study areas are based on agriculture (mainly rice and *rabi* crop cultivation), farmers are now gradually moving towards supari palm-based agroforestry. Relatively high land, which is not suitable for rice cultivation, is selected for the agroforestry system. A small drain is dug around the boundary and earth is spread over the field to make it even and raise it above normal flood level. Organic matter like cow dung is applied during land preparation.

Seedlings or saplings are used for planting. Most of the farmers raise seedlings on their homesteads and a few collect from nearby nurseries or from markets. For raising seedlings, a nursery bed (no specified size) is prepared in a shady place. A little soil work by mixing decomposed cow dung is done to prepare the bed, which is then leveled uniformly. Ripe fruit (nuts) are harvested from 20- to 25-year-old trees in March and April. The fresh fruit is sown on the prepared beds; nuts are deposited 12 to 15 cm below the soil and 15 cm apart. The nursery bed is covered with wet straw or palm leaves until the seeds germinate in July or August. Straw is also used to prevent evaporation of soil moisture and to control weeds in nursery beds. The seedlings are kept two to three years in nursery beds, and then transplanted on prepared land.

To provide the partial shade requires for initial growth of supari palms, farmers cultivate bananas on the prepared land before planting supari palm saplings. Banana saplings are planted 2.5 m apart, with 3.6 to 4.5 m between rows. After one or two years, the banana plants attain a height suitable to provide partial shade for supari palm saplings, which are transplanted to well prepared pits with 2.0 to 2.5 m between plants and 3.6 to 4.5 m between rows. Circular pits are dug with a radius of 15 cm and depth of 45 cm just near the bases of banana plants. Pits are filled with soil and cow dung in equal proportions. Transplanting is usually done in June or July. Banana plants are kept through their fruiting life, generally three or four years. After harvesting the bananas, the plants are cleared.

No pruning is required for supari palms. Farmers keep the agroforestry plots clean by weeding and climber cutting. Two or three weeding are done each year based on weed infestation. Weeds are used as mulch at the base of palms. Climbers other than betel vines are cut down once in year. Diseased, crooked, or dead palms are removed from plantations whenever necessary, and replanting is done in the following planting season. In general, farmers do not apply inorganic fertilizer or water at all for supari palms. They do some soil dressing at the bases of palms before the start of fruiting. The earth around the base of each palm is dug lightly by spade and left at the base of each palm. Farmers reported that this soil dressing influences the growth of tap roots, increases productivity, and prevent the exposure of roots by heavy rains. Leaves and green manure are used frequently in plantations.

Supari palms do not need intensive care or maintenance. But sometimes plantation productivity is reduced to some extent due to several kinds of pests and diseases. The enemies of this palm, according to farmers, include a beetle (*Dryctes rhinoceros*) and windstorms. The beetle is a scourge to all palms. It bores into the trunk and feeds on the tender heartleaves. When the terminal bud is damaged, the tree dies. There is no way to stop it, and farmers replace the dead trees by planting three- to four-year-old saplings. To protect the plantation from strong wind, farmers grow bamboo groves on the northwest sides of plots.

Termites often injure palms by eating the rootlets (new small roots). Farmers destroy termites by hand to control them. Sometimes woodpeckers make holes in the trunks of the palms, and strong wind may break their trunks. Some planted saplings are damaged or killed due to faulty handling or to damages caused by cowboys and grazing animals. Farmers usually plant hedges of lemon, mehadi-kata (*Pithecellobium dulce*), mandar (*Erythrina* spp.), or other trees around agroforestry plots to keep out grazing animals.

A disease locally called *tepa* attacks the top of the stem and forms an unusual canker structure. Farmers mentioned that if the disease is not controlled quickly, the tree may die within two or three years. The farmers developed a technique to control this disease. They make a slanted cut at the side of the *tepa* (enlarged canker structure) using a *dao* (a kind of big knife) so that gum may ooze from the affected portion. Farmers reported that this method satisfactorily prevents the disease. Recently many old palms suffer damage from an unidentified disease. In this case, severe top dying of old trees is observed. Farmers remove the affected trees from their farms. Some farmers use *hookah pani* (water from a traditional smoking devices) to check the disease at the initial stage of top dying, though the result is not satisfactory. Another disease called *kalia* affects the immature fruit. To prevent this, farmers use kitchen ash.

Nuts (*supari*) are the main product of supari palms. Generally, the palm bears fruit at seven or eight years of age. The tree flowers in February or March and the fruit is ready for harvesting from September through December. Farmers harvest in two phases: immature and ripe. They use a thin bamboo pole to which a *kachte* (a knife-like device used to harvest rice) is attached to cut the bunches of nuts. This technique does not damage the betel vines that climb on supari palms.

When the palms in agroforestry plots start to bear fruit for the first time, a second cycle of transplanting takes place just near the base of previous plants. Similarly, when the second generation is fruiting, a third generation is transplanted, and so on. As a result, a regular age gradation is maintained. When palms becomes old (generally 25 years old), they are felled and used for farmhouse construction or sold in markets.

Management of Betel

Betel is the main agricultural component in the supari palm-based agroforestry system. Planting of betel cuttings starts when supari palms attain a suitable height (usually seven or eight years after planting) to support the vines. Before planting, plantations are completely weeded. Farmers dig a very small pit just near the base of each palm. This land preparation is done from November through February and after the first shower in the pre-monsoon (April and May) planting starts. Leafy rooted cuttings (three or four nodes) locally called *powa* are used for planting. Cuttings are collected from homestead agroforestry plots. Planting one cutting per pit is general practice, but farmers plant two or three cuttings to compensate for any losses. When the vines twine on palms, farmers use large sheaves of leaves to tie the creepers onto the palms.

Grounds are kept clean by weeding. Weeds are used as mulch. Straw and nut husk also use as mulch at the base cuttings. No inorganic fertilizers are used. Farmers reported that insect infestation on vines is very rare. A type of black ant sometimes damages the nodes of young vines. In recent years, farmers observed diseases including those called *kalia* and *lalpocha*. These diseases rot betel leaves beginning at the outer edges moving inward. At the primary stage of infection, farmers often use ash as a preventive, but the result is not satisfactory. In cases of severe infection, vines are uprooted and buried.

Newly planted cuttings become ready for leaf production at two or three years of age. Farmers do the first picking in third year and continue for 12 years. After seven years, leaf production declines. Leaves are harvested twice a year during and after the monsoon. Selective picking is done during dry months. Normally monsoon harvests yield more leaves than dry month harvests. Leaves are harvested so as to not hamper the future growth of the vines.

Management of Other Perennials and Annual Crops

Pineapple is one of the important components of this agroforestry system. Farmers allocate some portion of their agroforestry plots to pineapple cultivation. They prepare land for pineapple planting during dry months (January to March). Two ploughs are needed to prepare land, which is then leveled. Care is taken not to damage betel vines and supari palms during ploughing.

Cow dung, urea, and TSP are applied at the time of field preparation. No uniform ratio[2] is followed in applying fertilizers. Pineapple suckers, locally called *potha,* are planted in rows during the months of January through April. Distance between rows is about 60 cm and that between plants is 45 cm. After planting *potha,* the whole field is covered with straw. Sometimes, leaves of *mandar* and other trees are used as mulch. These materials decompose within one year. The area is kept free of weeds. NPK fertilizers are applied every year in a ratio of 1:1:1; top dressing of soil is done for applying fertilizers. Fields are irrigated in the first year a few days after planting.

Pineapple begins fruiting about 18 months after planting. A pineapple bush has two flushes of fruiting. The first flush takes place during the months of May through July. The fruit is locally called *auguri phal.* The second flush in the months of December through February and the fruit is called *nomola phal.* The third year is the peak fruiting season of pineapple. After that farmers keep pineapple plants for another two or three years, or raise a new plantation in same place.

Besides pineapple, farmers cultivate other agricultural crops like turmeric, ginger, arum, and sweet potatoes based on family needs and market demand. When supari palms become long enough to permit partial sunlight to reach the ground, farmers grow these crops. Most farmers practice mixed cropping. Farmers take intensive care of these crops as they provide a handsome seasonal income. They apply fertilizers and irrigate the crops whenever necessary.

Farmers also reported that they plant some edible fruit trees like litchu (*Litchi chinensis*), peyara (*Psidium guajava*), ata (*Annona squamosa*), kamranga (*Averrohoa carambola*), and latkan (*Bixa orellana*) as boundary plantation around agroforestry plots. The fruit is mainly consumed by household members, though some farmers sell a portion in markets.

Production and Income from Agroforestry Products

A number of products are harvested from agroforestry plots year-round. The main marketable products are nuts (*supari*), betel leaves, pineapples, and other agricultural products.

Though palms bear fruit from seven years of age, in plantations they rarely fruit before 10 or 12 years. Farmers maintain different age gradations to maintain production continuity. Nut productivity varies considerably with tree age, weather conditions, and site quality. However, it was observed that the average annual fruit production (number of nuts) per tree ranges from 110 to 120. Even though market value is less, farmers harvest a small number of nuts at an immature stage to get immediate subsistence income. Poor farmers sell a larger amount of nuts at an immature stage. Farmers face economic crises from May to September or October almost every year due to a lack of off-farm jobs. This crisis compels them to sell immature nuts at a relatively cheaper price. Most of the farmers sell their products as mature nuts. Prices of mature nuts are higher than those of immature and *maza nuts*. However, farmers of the medium and large categories preserve almost 50 per cent of nuts as *maza*. To make *maza*, the ripe fruit are kept in a large earthen container filled with water and submerged. In this way they can preserve nuts for more than one year. These *maza* nuts are priced high and generally marketed when immature and mature nuts are not available. Farmers sell nuts by the *kahon* (1 *kahon* = 1280 nuts). Very small portions of nuts (usually dried nuts) are also consumed by households. During the study period, the mean market prices of different palm products (immature, mature, and *maza* nuts) given by farmers were Tk.(Bangladesh currency-Taka) 425, Tk. 825, and Tk. 800 per *kahon*, respectively (Tk. 58.00 = US$1 in 2002).

Betel leaf in the study area is marketed in the form of *motha* and *bishi*. Normally a *motha* contains 16-20 individual leaves with leafstalks while each *bishi* consists of 1344 leaves. In case of *bishi*, initially 64 leaves are arranged into a set, called a *sha*. Twenty-one *sha* make a *bishi*. Leaves are marketed on a *motha* basis during *Baishak-Shravan* (May through August) and as *bishi* in *Agharayan-Chaitra* (December through April). The market price for each *motha* and *bishi* varied from Tk. 10 to 20 and Tk. 300 to 400, respectively.

Besides these nuts and betel leaf, farmers grow other intercrops like ginger, turmeric, arum, *mashu alu* (a kind of potato), bananas, sweet potatoes, and pineapple. Farmers consume major portions of these products, and sell the surpluses in the village markets. They generally do not keep any records of intercrop production, but they could remember how much money they received from the sale of products.

Though this agroforestry system needs only very simple maintenance, there are some costs involved with cultivation, tending and fertilization, labor, harvesting, and marketing of products. Farmers usually do not maintain any records, but they knew the total amounts of their expenses. Across household categories, the range of costs varied from Tk. 587.00 to Tk. 7792.00. Despite these costs, farmers earned a handsome profit from their agroforestry. The average profit in poor category was Tk. 4799.00, and it substantially increased in other farm categories.

We asked farmers to tell us their total family income to find the contribution of agroforestry in the household economy. Last year their family income was derived from the sale of surplus rice, other crops, agroforestry products, livestock, small village business, and other sources. Their accounts showed that besides providing various goods for household consumption, this traditional agroforestry system contributes on an average 16 per cent of total family income across all farming categories.

Marketing of Products

Farmers produce a number of marketable products from their agroforestry plots. These products benefit the farmers when they are sold in markets. There are village markets as well as large markets in the district towns. Marketing of products and marketing channels are very important for the farmers because a better and improved marketing system can enhance the economic conditions of the producers and encourage them to continue production (Ahmed and Islam, 1995).

Farmers sell their products at both village and district markets and maintain some channels between producers and consumers. Poor farmers usually sell their products directly to consumers in the village (or local) markets on market days (normally twice a week). Sometimes they vend the products in nearby villages, carrying them on their shoulders or heads. Rich farmers with a substantial portion of marketable products generally sell their products to middlemen. The local middlemen (*bepary*) generally purchase these products either from village/local markets or from farmers. They bring all products to the upazila/district markets and sell to the wholesalers (*aratdar*). Businessmen from different parts of the country come to these *arats* (wholesale shops) to buy and distribute some of these products throughout the country. *Aratdars* also sell different products to small resident and local wandering traders (*farias*), who in turn sell them to consumers.

We tried to find how the price of goods varied when they were traded by intermediaries in marketing channels. A market survey found that *beparies* sell products to *aratdars* at a profit of at least 10 to 15 per cent. *Aratdars* make a 5 to 10 per cent profit, and *farias* make a higher profit at around 10 to 20 per cent. When the products are distributed to other districts of the country, the price almost triples. This means that when products are sold to final consumers through intermediaries the final price increase is about 25 to 45 per cent of the original price paid to producers.

Constraints and Opportunities of the Production System

Constraints

Although the management of the agroforestry system seems simple and profitable, farmers face several constraints. The main problems in the study areas are shortage of land particularly for marginal and small farmers, crop loss due to pest and diseases, the high costs of fertilizer, distant markets and poor transport links, pilferage of fruit, credit facilities, damage by natural calamities, and above all, lack of institutional support. Farmers can take remedial measures for some pests and diseases, but problems still remain with some unknown pests and diseases. Sometimes they remove the damaged plants and fill the gap with three- to four-year-old saplings. Application of specific pesticides or selection of pest- and disease-resistant genotypes may overcome this problem. There is no easy solution for the shortage of land, although there are available government lands (*e.g.*, roadsides, canal banks, etc.) that could be distributed among Poor and small farmers with usage rights. Farmers usually sell their products in village markets. The price in village markets is generally lower than in sub-district or district markets. But transport links with district markets are not good and require high transport costs. Cooperative marketing systems can be developed to get higher prices in district markets. Pilferage of ripe fruit is a great problem in the study areas. Plots located at a distance from a farmer's residence are vulnerable to this problem. Farmers reported that at least 25 per cent of the ripe nuts are pilfered from these plots. They sometimes hire sentries at night to protect nuts from thieves. Farmers claim that the agroforestry system provides them with livelihood support, but the government pays no attention. They believe that if government institutions like the Forest and Agricultural Extension Departments would help them with crop improvement, protection from pest and diseases, and marketing of products, they would benefit more from this agroforestry system.

Opportunities

The cultivation of supari palms with other crops has been found to be a popular and common feature among farmers of the study areas. During field data collection, we asked not only the sampled farmers, but also small traders, day laborers, and elderly villagers, about the opportunities afforded by the supari palm-based agroforestry system. All people responded that supari palms could grow on almost all types of land including farmland, fallow land, homesteads, and elevated land. Thus unproductive (unsuitable for agriculture) land can be brought under palm-based agroforestry. Once seedlings are planted, they need very little care and remain productive for a long period (25-30 years). Betel leaf is also productive for 10 to 15 years. Long productivity and little maintenance encourage farmers to practice this multiple land use system. They take care the *supari bagan* (supari palm plantations) generally in the winter season when they are free from other agricultural chores. Market demand for products (nuts and betel leaves) is another incentive to practice this agroforestry system. Farmers believe that this type of agroforestry system could be established in other suitable parts of the country.

Discussion

Traditionally people of rural Bangladesh grow trees, shrubs, and vegetables in a mixture on homesteads and also on croplands, based on their manifold needs and generation-old experience. The arrangement of different tree species around the house, taking into account the needs for windbreaks, shade, and open space, as well as the distribution of different shrubs and annual vegetables in different micro-niches, seem quite appropriate. The practice of traditional agroforestry provides them with additional income and meets a considerable portion of household needs (Abedin and Quddus 1991; Nath *et al.*, 2002). Availability of land, market demand,

and household needs encourage them to carry out this type of agroforestry system.

Differences Among Farmers in Allocation of Land, Tree Density, Income, and Marketing of Products

Among farmers of the studied households, considerable differences have been observed regarding land allocation for agroforestry, stocks of supari palm on agroforestry plots, marketing of products, and the income derived from the sale of agroforestry products. Farmers mentioned that sustainable financial incentives, long-term productivity, and multiple products encouraged them to practice agroforestry on land which had become unsuitable for rice production due to adverse weather. Viswanath *et al.* (2000) also mentioned that ease of management, multiple products and services, and access to markets encouraged farmers to adopt traditional agroforestry systems. Poor farmers are more vulnerable to frequent crop loss as their livelihoods depend on small holdings. For better use and productivity, they allocated more land to agroforestry than other farmers. Hocking and Islam (1998) observed that though it reduces rice yield, farmers kept trees on cropland. Farmers compensate rice loss by the production of wood, fodder, and fruit from the trees. However, other farmers argued that as agroforestry requires continuous supervision, and because they have more farmland and other business to support themselves, they obtained little income from agroforestry. Drawing on Nair and Krishnankutty (1984), Nair and Sreedharan (1989) reported that cropping intensity in terms of total area is higher in smallholders than large farmers. This indicates poor farmers utilized their land to the fullest extent to sustain their livelihoods.

Considerable variation among farmers was also observed regarding the care of trees. This variation was more obvious when tree density was measured. The number of palms per ha in the Poor category was the highest. The main reasons for the difference in tree density, as we observed and farmers claimed, was the nature of care and the distance from home. Smallholders' agroforestry plots are located near their homesteads where women and other family members have easy access to care for the trees. Women play a significant role in agroforestry management in the study areas. On the other hand, large landholders developed agroforestry plots somewhat remote from their homesteads. Tree mortality rates in distant plots were higher due to lack of proper care. Good management of agroforestry plots by Poor farmers was also manifested when we investigated the income per ha of agroforestry plots. It was observed that Poor families had the highest income per ha. These variations were due to the intensive care of agroforestry plots by poor farmers. Farmers reported that regular monitoring of plantations is essential. It helps them quickly identify any disease or pest outbreaks and nutrient deficiencies, allowing them to be controlled with very little effort. Poor farmers believe that existing land is all they will have in life, and that it comprises their livelihood assets. Even though income per ha was highest, the agroforestry system contributes only 12 per cent of family income for Poor families in comparison to 19 per cent for larger landholders (see endnote 3). However, these variations in income among farmers occurred because of land area allotted for agroforestry, *i.e.*, the more land allotted to the agroforestry system, the more income derived from it. The poor families are satisfied with small contribution because they have very limited land and can use their land in a sustainable and economic manner.

These findings are important from a development perspective. In Bangladesh, particularly in northern parts of the country, promotion of agroforestry under the intervention of the social forestry program began in the early 1990s. The program, however, failed to address the needs of the poor as most of the benefits were siphoned off by social elites at the expense of Poor people. On behalf of poor farmers, influential people who have a good mutual understating with project officials became the project beneficiaries. Subsequently, the raised plantations become unsuccessful due to lack of proper care. On the other hand, it can be assumed that if poor people were directly connected with project, it would be a successful one. Poor farmers have more interest in production systems which support their livelihoods.

The findings of tree density are consistent with Hocking *et al.* (1996). They find that though larger and wealthier farmers plant more trees, the survival rate is lower than that of plantations worked by poor farmers. The reasons include: 1] in the case of poor farmers, both men and women select species, and women have a direct interest in and responsibility for looking after the trees, and 2] larger farmers select species without consultation with women, and recruit employees to care for the planted trees. Employees appeared to have less of a stake in the success of the trees and consequently look less care of them. Nair and Sreedharan (1989) and Viswanath *et al.* (2000) also found that tree density decreased as landholdings increased. The latter mentioned that due to low investment and maintenance costs, and the multiple productivity of the system, poor farmers are more interested in taking care of trees, and therefore adopt traditional agroforestry systems. In contrast, Price and Campbell (1998) observed that in homesteads of Zimbabwe wealth made little difference to the density of trees planted and conserved. Moreover, Coomes and Burt (1997) reported that larger landholdings tend to use potentially more sustainable as well as more lucrative agroforestry systems.

Farmers sell a portion of their nut harvests while still immature, even though the nuts would bring higher prices if sold at maturity. To alleviate the financial crisis during the idle period (May to October), the government and NGOs may take initiatives to introduce small-scale enterprises. NGOs through their credit programs may encourage farmers to set up cottage industries. In villages, particularly villages away from town, there is a great demand for cottage-industry products such as baskets and agricultural implements. To increase their benefit, Poor farmers sell their products directly to consumers in markets. In larger farm categories, farmers sell their products through middlemen to avoid problems in transporting products to markets.

Adoption of Supari Palm-Based Agroforestry System

Preferences, resources, market incentives, biophysical factors, and uncertainty are the five factor clusters influencing the adoption of technologies such as agroforestry (Pattanayak *et al.*, 2003). Different resource endowments encourage different types of agroforestry practices. Resource endowments may include asset holdings such as land, labor, livestock, and savings. Generally resource endowments are likely to be positively correlated with the probability of agroforestry adoption. Market incentives focus on the explicitly economic determinants of adoption such as prices, market availability, transportation, and potential income losses or gains. Biophysical factors relate to influences on the physical production process associated with farming and forestry. Examples include soil quality, slope of farmland, and plot size. In general, poorer biophysical production conditions (*e.g.*, greater slope or potential for high erosion) create a positive incentive to adopt technologies that will mitigate these conditions. However, it is also possible that some farms are of a quality that is below the threshold of useful investment. The multiple-output, multiple-input, multi-seasonal character of agroforestry makes it a potential mechanism for reducing risk and uncertainty by diversifying the farmer's portfolio, and is therefore a good candidate for adoption by small holders.

Farmers in the study areas have both homestead and farmland. Due to adverse climatic conditions (*e.g.*, long droughts and flooding) farmers face difficulty in rice cultivation. This trend of frequent crop loss encourages farmers to change their farming practices. It was observed that poor farmers are more vulnerable to repeated crop loss as they have a very limited amount of farmland. This trend toward a decline in rice production encouraged them to adopt agroforestry more readily than did rich farmers. They also mentioned that there is less chance of crop loss and a considerably long period of sustainable production in supari palm-based agroforestry. By contrast, larger farmers can minimize crop loss as they have more land for *boro* rice cultivation in the winter season. *Boro* rice cultivation is costly because it needs irrigation and more fertilization, which small farmers cannot afford. However, only land unsuitable for rice cultivation (generally raised land) is allotted for agroforestry because this land drains well in monsoons and cannot be irrigated for *boro* rice farming in the late winter season. The market for agroforestry products (mainly betel leaf and betel nut) is buoyant. The leaf is usually chewed with slices of betel nut and lime. It is a common and sometimes mandatory item on many social occasions. These products have a huge internal market demand, and the market for betel leaf has expanded in foreign countries as well (Nath *et al.*, 2003). A secured market is considered one of the important factors for sustaining any farming system (Deomampo 1994). As there are internal as well as foreign markets for the products of this agroforestry system, the future of the system appears promising. Maintenance of regular age gradations also helps sustainable production. Besides producing leaves and nuts, a number of other agricultural crops are also produced in agroforestry plots. Surplus products are traded in local markets. As the system accommodates several crops that are harvested in different time periods, the agroforestry system provides income almost year-round.

Crop Interactions

Exploitation of interactions between woody and non-woody (*e.g.*, annual crops) components is the key to the success of all agroforestry systems (Rao *et al.* 1998). A better understanding of the interactions provides strong scientific bases for improvement of traditional systems. In study areas, rural people are seen to cultivate deliberately different crops including trees, herbs, shrubs, and grasses on a single plot of land. The management of different crops is part and parcel of their land use strategy. They are well acquainted with crop combinations in traditional agroforestry systems. This crop combination strategy can be explained in terms of the farmers' indigenous ecological knowledge about plant interactions. When they were asked to explain the crop interactions, they shed light on the relative height, crown structure and sunlight requirements of different species. Farmers have more knowledge about above-ground interactions than below-ground interactions. This is probably simply because above-ground interactions are more readily observed. They mentioned that from the success of neighbors and experiences of the elders they have been able to find out what types of plant can grow well in what conditions. For example, pineapples, turmeric, ginger, and other crops grow well in partial shade; betel needs support trees and partial shade; and betel nuts require partial shade in the initial growth stage. Based on this knowledge they cultivate diverse crops in the different growth stages of plants. To prevent excessive shade, they maintain adequate space between plants. The sparse canopy of the scattered trees is unlikely to cause any serious shading of the undercrop at tree densities practiced by the farmers in traditional agroforestry systems (Viswanath *et al.*, 2000).

The sharing of moisture and nutrients among different crops is also important as it enhances growth and better utilization of nutrients. Farmers mentioned that after a few days of fertilizing and irrigating agricultural crops, betel leaf and betel nut plants become greener, which signifies that these plants benefit from fertilization and irrigation. This is consistent with the findings of Nissen *et al.* (1999). They mentioned that intercropping of well-fertilized annuals with trees will benefit the trees from intensive fertilization and weed management of the intercrops. Among the farmers, variation was observed in the proportion of fertilizer application. Rich farmers apply more manure (cow dung) and TSP than poor farmers. Though we did not investigate the variation in productivity due to fertilizer application, it needs exploration to find an optimum level of nutrient supply for financially better production.

Farmers' ecological knowledge about plant interactions appears to be comparable to scientific knowledge (Teklehaimanot *et al.*, 2001). Farmers were able to recognize crown size and density as attributes of trees that affect shade. According to Teketay and Tegineh (1991), tree shade ameliorates the microclimate of lower story crops by reducing air and soil temperature.

The supari palm is the predominant crop of home gardens in southern coastal Bangladesh where it is widely grown as a cash crop both along the boundaries of home gardens and as block plantings within them (Nath and Karmakar, 2001). In certain areas, it is cultivated in orchards. However, farmers in these areas grow the palm as a single crop; no other crops are integrated with palms. The opportunity to introduce other crops like betel with supari palms should be investigated. Moreover, there are vast areas of degraded and encroached forest land, and public and private fallow land, where this type of agroforestry system could be replicated with the active participation of local people.

Farmers of rural Bangladesh have practiced different types of traditional agroforestry systems for decades. These indigenous agroforestry systems provide one of highest returns on investment of any low-input production option for the tropics (Teklehaimanot *et al.*, 2001). The indigenous management techniques for age-old agroforestry and the ecological knowledge of the farmers operate at an effective but simple level. This knowledge could be incorporated with scientific understanding for further improvement of the agroforestry system.

Conclusions

Indigenous agroforestry systems are found to provide subsistence and cash income while conserving soil, water, and forest resources. These systems are considered to be socially appropriate because they are practiced by smallholders who do not have the resources to adopt high-input agriculture or practice the agency-recommended systems (Coomes and Burt 1997; Viswanath *et al.* 2000). Though smallholders' agroforestry contributes substantially to rural livelihoods and national economies, this contribution is not adequately quantified or appreciated (Garrity 2004). Insight into why local farmers engage in distinct practices is essential not only to inform research and field extension efforts targeted at particular groups, but also to help the broader quest of identifying the conditions and circumstances under which agroforestry offers a viable and alternative path for agricultural development (Coomes and Burt 1997).

Diverse types of traditional agroforestry systems have evolved in different parts of Bangladesh based on farmers' needs, market incentives, and physical factors like climate and soil. The physical factors in the study areas are suitable for the growth and yield of *supari*-based agroforestry components. Farmers are well equipped with skills and knowledge for the sustainable management of their agroforestry. The agroforestry system provides off-farm income. It also generates income sources for other people engaged in trading, transportation, and other occupations. Frequent crop (rice) loss, secured markets, and sustainable and long-term production periods are the factors that encourage poor farmers to adopt the supari palm-based agroforestry system. Small farmers seem more efficient in the management of their agroforestry plots, as seen in tree density and production per hectare. Along with the many positive aspects, however, farmers face some problems related with management and marketing of agroforestry products. Necessary institutional support both from the government and non-governmental organizations is needed to solve these problems.

Several earlier studies (*e.g.*, Nath and Karmakar 2001; Nath *et al.* 2003) reported that supari palms are able to grow in diverse site conditions. The possibility of introducing betel and other suitable agricultural crops with supari palms in other parts of the country deserves further exploration.

Notes

1. Farmers in the study areas are used to cultivating rice on farm land. But due to long flooding during monsoons and drought in the summer, they experienced crop loss almost every year. To compensate for this crop loss, they developed

supari palm-based agroforestry on relatively raised farm land.

2. Those farmers having large amounts of cow dung apply more than other farmers. Regarding urea and TSP, rich farmers follow a 1:1 or sometimes a 1:2 ratio, whereas poor farmers mostly use a 1:0.5 ratio. Rich farmers believed that though TSP is comparatively expensive, its effects on plants continued a long time.
3. Though farmers could recall their gross total family income per year, their figures seemed to be quite rough. Income disparity among and also within farmer categories was very high. However, we calculated the contribution of agroforestry to total family income, because even rough data would help us gain a general understanding.

References

Abedin M. Z. and Quddus M. A. 1991. Agroforestry systems in Bangladesh with particular reference to economics and tenurial issues. In: Mellink W., Rao Y. S. and MacDicken K. G. (eds), *Agroforestry in Asia and Pacific*. RAPA publication: 1991/5, Bangkok. pages 13-33.

Ahmed F. U. 1995. Traditional agroforestry systems in Bangladesh. Agrofor Newsletter, 3: 6-9.

Ahmed M.R. and Islam A. 1995. Marketing pattern of some multipurpose tree species production in Bangladesh. Bangladesh J of For Sci 24(2): 41-48.

Bhosale D. B., Thakare R. P. and Deorukhkar A.C. 1994. Comparative economics of spices (black pepper and nutmeg) as intercrops in south Konkan (Maharastra State). South Ind Horti 42: 224-230.

Coomes O. T. and Burt G. J. 1997. Indigenous market-oriented agroforestry: dissecting local diversity in western Amazonia. Agrofor Syst 37: 27-44.

Deomampo N.R. 1994. Market-lead farming system's approach for sustainable development in the hill and mountain areas. Paper presented at an International Workshop on Evolution of Hill and Mountain Farming Systems held at Lumle Agricultural

Research Centre, Nepal from 3-6 October, 1994 jointly organized by ICIMOD and FAO.

FAO (Food and Agriculture Organization). 1988. Climatic data base. Technical Report-3(1). UNDP/FAO, BGD/81/1035, Dhaka.

FAO/UNDP. 1988. Land resources appraisal of Bangladesh for agricultural development. Report 2: Agroecological regions of Bangladesh.

Garrity D. P. 2004. Agroforestry and the achievement of the Millennium Development Goals. Agrofor Syst 61: 5-17.

Gujral R. S. 1991. Types of agroforestry systems in Asia and Pacific regions. In: Mellink W., Rao Y. S. and MacDicken K. G. (eds), Agroforestry in Asia and Pacific. RAPA publication: 1991/5, Bangkok. Pages 195-201.

Hammermaster E. T. 1981. Project report on inventory results for village forest inventory of Bangladesh. UNDP/FAO project, BGD/78/020, Field Document No. 5. Dhaka.

Hocking D., Hocking A. and Islam K. 1996. Trees on farms in Bangladesh 3. farmers' species performances for homestead trees, survival of new tree planting, and main causes of tree death. Agrofor Syst 33: 231-247.

Hocking D., Sarwar G. and Yousuf S. A. 1997. Trees on farms in Bangladesh 4. Crop yields underneath traditionally managed mature trees. Agrofor Syst 35: 1-13.

Hocking D. and Islam, K. 1998. Trees on farms in Bangladesh: 5. growth of top- and root-pruned trees in wetland rice fields and yields of understorey crops. Agrofor Syst 39: 101-115.

Hussain M. S. 1992. Soil classification with special reference to the soils of Bangladesh. University of Dhaka, Bangladesh.

Jacob V. J. and Alles W. S. 1987. Kandayan gardens of Sri Lanka. Agrofor Syst 5 (2): 123-137.

Korikanthimath V. S. and Rajendra H. 1994. Cardamom and areca nut mix-cropping systems. Ind Coca, Areca nut and Spices J 18: 109-112.

Leuschner W. A. and Khaleque K. 1987. Homestead agroforestry in Bangladesh. Agrofor Syst 5 (2): 139-151.

Liyanage M. de S., Tejwani K. G. and Nair P.K.R. 1989. Intercropping under coconuts in Sri Lanka. In: Nair P. K. R. (ed), Agroforestry Systems in the Tropics. Kluwer Academic Publishers and ICRAF, Netherlands. Pages 165-179.

Mustafa M., Hall J. B. and Teklehaimanot Z. 2000. Indigenous management techniques in Bangladesh homegardens. Int Tree Crop J 10: 215-228.

Nair C. T. S. and Krishnankutty C. N. 1984. Socio-economic factors influencing farm forestry: a case study of tree cropping in the homesteads in Kerala, India. Paper presented at the EWC/FAO Workshop on Socio-economic aspects of social forestry in the Asia-Pacific region. Bangkok, Thailand.

Nair M. A. and Sreedharan C. 1989. Agroforestry farming systems in the homesteads of Kerala, southern India. In: Nair P. K. R. (ed), Agroforestry Systems in the Tropics. Kluwer Academic Publishers and ICRAF, Netherlands. Pages 139-163.

Nath T. K. and Karmakar N. C. 2001. Traditional farming system of supari (*Areca catechu*) in rural Bangladesh. PALMS 45(3): 109-113.

Nath T. K., Karmakar N. C., Molla M. A. H. and Kabir M. A. 2002. Tal (*Borassus flabellifer* Linn.: Areacaceae): a valuable economic palm of rural Bangladesh. Forests, Trees and Livelihoods 12 (4): 283-295.

Nath T. K., Inoue M., Islam M. J. and Kabir M. A. 2003. The Khasia tribe of northeastern Bangladesh: their socio-economic status, hill farming practices and impacts on forest conservation. Forests, Trees and Livelihoods 13(4): 297-311.

Nissen T. M., Midmore D. J. and Cabrera M. L. 1999. Aboveground and belowground competition between intercropped cabbage and young *Eucalyptus torelliana*. Agrofor Syst 46: 83-93.

Pattanayak S. K., Mercer D. E., Sills E. and Yang J. C. 2003. Taking stock of agroforestry adoption studies. Agrofor Syst 57: 173-186.

Price L. and Campbell B. 1998. Household tree holdings: a case study in Mutoko communal area, Zimbabwe. Agrofor Syst 39: 205-210.

Rao M. R., Nair P. K. R. and Ong, C. K. 1998. Biophysical interactions in tropical agroforestry systems. Agrofor Syst 38: 3-50.

Teketay D. and Tegineh A. 1991. Shade trees of coffee in Harerge, eastern Ethiopia. Int Tree Crops J 7: 17-27.

Teklehaimanot Z., Millat-e-Mustafa M. and Hall J. B. 2001. Farmers' ecological knowledge about homegarden production systems in Bangladesh. Forests, Trees and Livelihoods 11: 67-79.

Viswanath S., Nair P. K. R., Kaushik P. K. and Prakasam U. 2000. *Acacia nilotica* in rice fields: a traditional agroforestry system in central India. Agrofor Syst 50: 157-177.

Chapter 5

Farmers' Management of Natural Resources for Livelihoods in the Mountains:

Valuing Traditional Knowledge and Strategies in the Hindu Kush-Himalayan Region

☆ *Vir Singh and Neelendra K. Joshi*

"The traditional practices of agro-ecosystem management that have evolved over 5,000 years of trial and error are a set of practices and traditions that should be analysed and understood rather than condemned as an unwelcome remnant of the past. Such understanding is an essential part of appreciating our human heritage. Traditional practices may provide the best base on which to develop effective production systems with no or minimum fuel subsidies, and survival of human population depends on the evolution of solar-powered agro-ecosystems that are stable and more productive".

–Mitchell (1979)

"Seeing agriculture from the eyes of the peasant has the virtue of getting us closure to truth".

–Vallianatos (1994)

"The direction of science and technology has been to find new ways and means to exploit nature. This has resulted in the depletion of resources, both renewable and non-renewable, due to overexploitation to satisfy the insatiable demands of society. This development created temporary prosperity, but has also created pollution, war and poverty".

–Bahuguna (1995)

The Hindu Kush-Himalayan (HKH) Region is strikingly different from the other ecosystems on Earth. Comprising the world's youngest, highest, the most fragile and the most vulnerable mountains, the HKH Region itself is one of the hotspot areas torn by national and international conflicts, wars and natural calamities. Extending into the boundaries of eight Asian countries – Afghanistan, Pakistan, India, China, Nepal, Bhutan, Bangladesh and Myanmar – the HKH mountains are inhabited by some of the world's poorest communities. Economy of the Region is dependent on agriculture. About 70-90 per cent of the people in the mountains derive their livelihoods from land-based activities, viz., cropping, forestry, livestock husbandry, and pastoralism. Farming communities, in addition to being poor, are also one of the most marginalised. The inaccessibility-ridden Region is also rich in natural biodiversity, human cultures and traditional wisdom.

The land-based economy, or the agrarian economy, is driven by the traditional knowledge people are especially pregnant with. With the advent of the Green Revolution era in mid-1970s some of the more accessible and favourable areas of the HKH mountains could not refrain from witnessing agricultural transformation with the introduction of the so-called high-yielding varieties (HYVs). Ever since the process has been strengthened steadily and has been drastically affecting the farming cultures. Traditional wisdom in this paradigm of development has hardly any role to play.

The outcome of the undergoing transformation processes driven by external experts, external inputs and external knowledge system is dismal. Virtually the entire HKH Region is in hot waters today. Never before in the history have mountains been the subject of such a hot discussion the world over. The gloomy spectre is emerging owing to the negative implications of the agricultural transformation processes that not only dictate land use changes but also the traditional socio-economic framework of the communities in the Region. The negative changes, however, can be reversed and the processes of un-sustainability can be switched off for turning them towards sustainability. For this to make a reality we need to explore mountain farmers' traditional knowledge system and strategies built around it. This article attempts to present many facets of agriculture mountain farmers have evolved over millennia through trial and error.

Farmers' Perception of the Green Revolution

"These (modern) practices are adversely affecting the productivity of land and are destabilizing the ecological balance. They are worsening the dependence of the Third World on the industrialized countries for knowledge, techniques and inputs in areas where indigenous knowledge and resources are capable of, and were till recently, meeting the needs in a satisfactory fashion. Moreover, logging and forest degradation has also caused tremendous soil erosion in the Third World countries. Agriculture and connected practices being the major way of life of the Third World societies there is an urgent need to defend the indigenous knowledge, practices and resources in these areas from the western onslaught. We therefore recommend that the Third World countries should make joint efforts for the restoration of the indigenous way of life in these countries".

–Sardar (1988)

The conventional agriculture technology – the Green Revolution technology – was based on the assumption that technology is a superior substitute for nature, and hence a means of producing limitless growth, unconstrained by nature's limits. However, the assumption of nature as a source of scarcity, and technology as a source of abundance, leads to the creation of technologies which create new scarcities in nature through ecological destruction. The reduction in availability of fertile land and genetic diversity of crops as a result of the Green Revolution practices indicates that at the ecological level, the Green Revolution produced scarcity, not abundance (Shiva 1991).

Mountain areas, thanks to their terrain conditions and high degree of inaccessibility, did not permit Green Revolution to rein throughout. It has been possible only in the irrigated valleys. These favourable valleys have witnessed almost complete genetic transformation of food crops, especially of rice and wheat. Some market-linked valleys have been transformed into vegetable belts. Some of the patronised high-altitude areas have been converted into fruit belts with apple as the only fruit species. Only the mid-altitude rainfed areas have continued to be traditional.

The agricultural transformation in the mountains was officially regarded as a panacea for food security in the food-deficient region. Replacing the traditional crops by market-oriented cash crops was, however, no idea to realise food self-sufficiency in the Region, but to increase food purchase power of the people. Mountain farmers' experiences of Green Revolution have been quite bitter. Having adopted the Green Revolution in favourable areas, they have only harvested crops of hunger, malnutrition, and discontent. Some of the things they would recount as their experience about the Green Revolution are as follows:

- ✰ Change in agricultural values;
- ✰ Narrow straw-grain ratios leading to fodder shortage to livestock and drudgery on women;
- ✰ High degree of vulnerability for crop damages due to high incidence of epidemics and unfavourable weather conditions;
- ✰ Pollution of the environment leading to health hazards;
- ✰ Knocking down of soil flora and fauna and increasing incidence of desertification;
- ✰ Poor nutritive value and less palatability of new food grains;
- ✰ Ceasing of seeds to be common property resources;
- ✰ No free exchange of germplasm within community;
- ✰ Decrease in collective responsibility;
- ✰ Breakdown of social sanctions on resource usage;
- ✰ Slackness in social cohesion;
- ✰ Dependence on expensive, often unaffordable, external inputs and technology; and
- ✰ High degree of risks.

Having experienced unsustainability and food insecurity associated with the monoculture-based Green Revolution, the mountain farmers are now opting for agrobiodiversity-based farming practices. Owing to the poor accessibility and other terrain features of the Region, many remote, isolated, and somewhat inaccessible mountain areas did not witness a genetic wipe out as the transformed ones did. Mountain farmers for many years have been collecting their lost seeds from these areas and re-introducing them in the transformed areas from where these had vanished thanks to the introduction of fertilizer-responsive high yielding varieties (HYVs). Deprived of their genetic wealth, many Green Revolution areas in the mountains of Uttarakhand and other parts of HKH region are once again boasting of their rich agrobiodiversity (Singh 1995, 1998b).

Gender-specific Technologies

"To say that women and nature are intimately associated is not to say anything revolutionary. After, all it was precisely just such an assumption that allowed the domination of both women and nature. The new insight provided by rural women in the Third World is that women and nature are associated not in passivity but in creativity and in the maintenance of life".

–Shiva (1989)

Most of the agricultural operations in mountain areas are run by women farmers, and therefore, most of the technologies have been designed and developed by women farmers. Women's technologies are founded on the ecological principles. They are site-specific and resource-centred. They are humane and futuristic. They are conservation-oriented and resilience-enhancing. They infuse life and sustainability in the whole agroecosystem.

Women technologies are based on a rich repository of traditional knowledge that stored in human brains. The traditional technologies evolved particularly by women have been tested over millennia by trials and errors and are perfectly fool proof. These technologies are site-specific and negotiate with terrain conditions. Therefore a great diversity of the technologies exists in mountain areas, like the environmental diversity and the biological diversity in the area. Some of the specific attributes of the gender-specific technologies operating in traditional setting in the mountains are as follows (Singh and Tulachan 2002a, 2002b):

- ✰ Maintainance and promotion of equity and cohesion amongst the people of community;
- ✰ Provision of benefits to the poor villagers and large number of the poorest households;
- ✰ Addressing women's practical needs;

- ✰ Ensuring low investment and high returns;
- ✰ Site-specificity and suitability to mountain agro-ecosystems;
- ✰ Feasibility, social acceptability, and fulfilling people's priorities; and
- ✰ Innovation-inducing.

Traditional mountain technologies are the result of women's innovativeness and are still innovation-inducing. They do not only link the present with the past but also guide us into future development of natural resource conservation-oriented agricultural systems.

CPR-centred Farming

"There is no technology or even philosophy which allows humans to escape this earth to have 'misery' over it. We will always have to eat, and so we will always have to depend upon our earth".

–Durhan (1979)

Mountain farmers throughout the Hindu Kush-Himalayan (HKH) Region possess the area of arable land that is much less than that could provide them food security and a stable livelihood base (Partap 1999; Singh and Tulachan 2001). One of the essential features of mountain farming systems is the large uncultivated area, also known as the support area, rendering enormous ecological and economic services to the Region. The ratio between uncultivated and cultivated area is wide enough, further signifying the existing and prospective role of the former compared to that of the latter. Vast areas under the uncultivated land are the common property resources (CPRs). Local farmers have access to and some defined rights on these commons.

Mountain farmers have evolved over centuries the farming systems and the strategies that are compatible with the specific resource features of the mountain ecosystems (Jodha *et al.* 1992; Singh 1998b). CPRs – mainly the community lands under forests and grasslands – have been central to these strategies. Marginal farmers in the HKH Region are amongst the poorest communities of the world. They are, however, not resource-poor per se. They have access to and defined rights on common property resources (CPRs) under uncultivated support areas. Mountain farmers, from this yardstick, can be regarded as resource-rich. Their current poverty and pitiable condition are mainly attributable to their resource base degradation and continued negligence of the CPRs.

CPRs are at the heart of mountain agriculture. CPR-farmland linkages infuse vitality in the whole system. The uncultivated CPR lands provide a natural subsidy to the farmlands via the agency of livestock. These provide fodder to livestock due to which no proportion of cultivated area in mountains has to be devoted to fodder cultivation. In other words, there is no man-livestock competition on marginal farms. Only what are non-consumable products for human beings (crop residues, weeds, etc.) to grow in the farmland are provided to livestock. Almost entire fuel wood is extracted from the CPR areas, in the absence of which livestock dung would be used as a fuel for cooking to the impoverishment of soil fertility.

Ecological niches full of exceptionally high-value biodiversity–including CPRs, and genetic resources in the farmlands–would form the basis for the sustainable growth of mountains' own economies. Mountains' biodiversity, indeed, can also contribute to national, regional, and global economies. Marginality- and inaccessibility-ridden mountain farming systems will remain one of the richest repositories of agrobiodiversity on Earth in future and thus will be in a state to extend services to humanity in the foreland mainstream areas that are losing production potential due to serious erosive effects of monocultures of the Green Revolution combined with non-contribution/ absence of forests/ CPRs.

CPR-farm Linkages

"My general feeling about the technology is that when you promote a technology, it is very important to understand the nature of technology, whose interests it is serving, and how it is good for the farmer, or for that matter, who the constituencies of the technology are. Unless you know very clearly about that, then eventually the technologies will not be very fruitful".

–Rahamani (2001)

The biophysical resource base–land, soil, water sources, forest ecosystems, and agrobiodiversity–is basic to the Land, soil, water sources, forest ecosystems, and agrobiodiversity – i.e. the bio-physical resource base–is the key to the approaches to sustainability of food production and livelihood systems in the HKH Mountains. The other criterion is the forest-cropland ratio that must be wide enough so that the agro-ecosystem receives environmental services imparted by the forest area. Livestock, cropland and human beings have strong organic bondage with the forest ecosystems. The size of the forest area therefore would determine the health of livestock, cropland and human beings. Living soil, one of the basic capitals for sustainable agriculture, is the product of forest ecosystems. Continuous biomass flow from the forest (bedding material in animal sheds and fodder being converted into manure, for example) enriches the soil ecosystem, a process that supports agrobiodiversity and manifests into crop production (Tulachan and Neupane 1999; Singh 2000). Forests also influence the flow of water. In fact, nature's hydrological cycle and thus water sources in the mountains are also largely a product of the forests. Water flow/ circulation within an ecosystem is critical to the sustenance of life and production.

Farming systems in mountain areas are characterised by very high degree of diversity which must be preserved and enhanced. Despite the existence of an appropriate forestland-cropland ratio in the mountains of HKH Region, the forestlands exhibit a sombre picture of ecology. On large areas, there is no forest cover, or the forest density is very thin. If we take up measures to expand cropland area at the expense of forestlands, it would only disturb the delicate ecological balance of the mountain ecosystems. An increase in the cropland area becomes a compulsion when the production approach is commodity-driven that needs isolation of cropland from common property resources. The Green Revolution-type agriculture strictly observes this situation. If the forest sector is augmented, it would lead to improved cropland conditions and increased productivity.

Forests do not just improve the lot of cropland and increase productivity. They themselves offer a variety of high quality foods. These uncultivated areas, in fact, have potential to provide more quality food per unit area on sustained basis than the cultivated land (Singh 1991). Mountain inhabitants have been depending on foods produced in the uncultivated areas. The forest-based foods include edible fruits, flowers, buds, seeds, vegetables, mushrooms, honey, etc. and people have been dependent on these foods for millennia. Efficient management of CPRs should be our mandate if we are to produce uncultivated foods. CPRs are a very rich source of foods (Table 5.1).

Table 5.1: Number of Plant Species Providing Food in the Mountain Ecosystems and the Plains' Agriculture

Food Item	*Mountain Ecosystems**		*Plains' Agriculture***
	CPRs	*Cropland*	
Fruits	24	18	6
Vegetables	14	32	10
Cereals, Millets and Pseudo cereals	–	12	4
Pulses	–	10	5
Oilseeds	–	7	4
Medicinal and Aromatic Plants	99	4	–
Fodder***	52	9	–
Total Species	189	92	33

Source: Singh (2001).

*: Figures are based on a survey of Henwal Valley in the Garhwal area of the Uttarakhand Mountains.

**: Figures are based on a survey of the plains adjoining the mountains of Garhwal.

***: Livestock convert this forest biomass into human food, *e.g.*, milk.

CPRs thus offer to provide the largest number of edible food products (189). Numbers of plant species being cultivated in the cropland/ cultivated area (92) in mountain agro-ecosystems are also more that the ones grown in plain agriculture (33). Mountain agro-ecosystems in general and CPRs in particular, thus, are especially rich in the biodiversity of specific consumptive use values.

Diversity-complexity is the foundation of agricultural sustainability. Modernisation of agriculture has weakened this foundation of Indian agriculture. This has changed the state of diversity-complexity into monoculture-simplicity, and sustainability into unsustainability. Plains' agriculture, generally devoid of the CPR component, is less diverse and simpler than that of mountain agriculture. CPRs with natural green mantle provide ecological stability to the agriculture. Its missing in the farming system, as in the transformed plain areas, makes the whole system vulnerable to damage. However, mountain agriculture has too undergone significant changes wherever its modernisation has taken place. We shall find it in the following sections and in the case studies of the Indian Central Himalayas.

Productivity of fruit trees that form dominant phytocommunity is also many times more than that of cereal crops. For instance, when the land is used to produce animal protein, an acre of land can yield a hundred kilograms of beef in one year. The same land will produce one to one-and-a-half tones of cereals, seven tones of fruits or 10 to 15 tonnes of walnuts. If trees are grown to give edible seeds for human beings and livestock, the same land can grow 15 to 20 tonnes, besides enhancing the fertility of the soil (Bahuguna 1987-88).

Plantation of food-yielding trees, shrubs and herbs in the CPR areas would enhance agro-biodiversity and the ecological stability of the farming system along with the food security of the marginal farmers depending on them. Food products obtainable from trees are generally more palatable and of higher nutritive value than those provided by cultivated crops. Moreover, unlike the foodgrains and vegetables to come from the cultivated land, the variety of foods occurring in the CPRs does not require cooking or any other energy-consuming processing. Cultivated foods, contrarily, require fuel wood extraction from the forest areas, which often is one of the major causes of the state of ecological imbalance in the area. Food obtainable from forest area, however, does not make any such demand. Furthermore, the management of uncultivated areas for food by introducing food-providing woody perennials is ecologically a more stable system of food production than mere the cultivation of cereal crops on cultivated land.

Harnessing Biodiversity

"Responsibility for life at large obliges us to support diversity, for diversity is not only the spice of life, it is also a precondition of vibrant and healthy life; moreover, a precondition of the meaningfulness and richness of our individual lives".

–Skolimowski (1995)

Monocultures are extremely vulnerable. If they are promoted they contribute to make the food production systems vulnerable and unsustainable. Agroecosystems with many different niches occupied by many different kinds of species, however, are very resilient and very stable. Resilience, sustainability and farmers' security go hand in hand. The marginal farmers of the mountains stick to the principle of diversity. They would manage as high number of useful plant species as can thrive in a particular farming situation. Farmers would also maintain functional diversity, which can be achieved through combining complementary plant and animal species in synergetic interactions, which is one of the means to enhance sustainability of the mountain agro-ecosystems.

The *Baranaaja* culture of the mountain areas is based on the "Diversity is prosperity" principle. *Baranaaja* literally means twelve foodgrains intercropped with finger millet as the base crop. Amaranth, buckwheat, kidney beans, horse gram, black soybean, black gram, green gram, cowpea, rice bean, adzuki bean, sorghum, and *cleome are* the main crops intermixed with finger millet. In some areas even groundnuts are also mixed with the base crop. *Baranaaja* is a unique example of cultivating biodiversity in mountain areas. Managing agro-biodiversity through *baranaaja* helps farmers reduce risks and increase the level of their food and nutrition security. This is also a way of managing nutrients in the mountain soils. *Baranaaja* crops occupy only the rainfed and otherwise the least fertile areas of cropland where other moisture-demanding crops would fail to grow. The local farmers, using complimentarity and synergistic associations amongst different food-yielding species, have evolved agro-biodiversity-centred agricultural practices; the wonderful *baranaaja* culture being its glaring example. Such agro-biodiversity management enhances sustainability of the agro-ecosystem and raises the food security level of the farming community.

Genetic diversity of traditional crop cultivars in the mountains is also unique. A variety of rice landraces, each with specific traits, are cultivated in Himalayan valleys. Traditional rice varieties, like *Hansraj, Ramjawan, Kanguri, Bangway, Gorakhpuri, Basmati, Thapachini, Jhumkya*, etc., thrive in lowland valley areas, while *Chawaria, Munjil, Jhailda, Lekmal, Kallao, Almunji, Chwatu*, etc., grow well in upland rainfed areas and at high altitudes. Productivity of rainfed varieties is comparable with those of irrigated ones. Such rare, hardy, and sturdy varieties would hardly be found in the plain areas anywhere in the world (Singh and Jardhari, 2000). These landraces are the products of farmers' millennia old experience of discovery, selection, breeding and utilisation.

Landraces capable of thriving in harsh environments do not only satisfy people's hunger and contribute to food security, but they are also used in many rituals. All through the history of Indian civilization, these landraces have been used as potent symbols of religion and culture. These can heal the environment the HYVs have deteriorated due to their insatiable requirement of chemical fertilizers and synthetic pesticides the use of which has not only severely affected human health, but has also disturbed nature's food chain, giving a blow to delicate ecological balance. Superb landraces, indeed, can be used as a warrant for the arrest of hunger and malnutrition. And with them as an indispensable component of agrobiodiversity-centred farming practices in the mountains, we can usher in a new era of agroecological renaissance (Singh 2005).

Mountain livelihoods are also characterised by livestock farming. Livestock play very crucial social, economic, ecological and cultural role in mountain areas. By transferring nutrients from forest ecosystem to farmlands and maintaining cyclic flow pattern in the farming system, they contribute to the ecological integrity of the system (Singh 1998a, 1998b, 2000; Tulachan 1998; Tulachan and Neupane 1999). Mountain communities, in essence, are the livestock-dependent communities.

High degree of biodiversity creates barriers against any natural calamity (drought, for example) and pest epidemics. Some living organisms may be host to certain insects or diseases, some may be little more prone to weather extremes or other adverse conditions, but a complex mix of naturally occurring (in natural forest areas) organisms will do away any chances of widespread loss to the farmers. Marginal farmers in the HKH region would cultivate many-many different varieties for different purposes (productivity, drought resistance, taste, colour, nutritive value, rituals, etc.) and in doing so they enhance the degree of their security.

Intensive Use of Farm Elements

"Nature is a complex mosaic of phenomena. When we observe it, we are unable to 'see' or 'take in', or comprehend it in its totality. We mentally focus on certain aspects, often to the complete neglect of other aspects".

–Jackson (2002)

Intensive use of farm products forms one of the key factors of resilience and sustainability. It ensures higher degree of complexity in the farming system. Complexity is created by using a farm element to its optimum effect through using it as fully and intensive as possible by way of internal cycling and recycling (Werf 1993). Mountain farmers would use a farm produce or element at several levels and in several functions, making waste/ intermediate product or function the input for another and ultimately cycling and recycling the end product into the farming system. A large number of elements in this way are placed into a web of symbiotic and dynamic interrelationships. An output at one level becomes an input at other level. Larger the number of functions fulfilled by a single farm elements, the higher the degree of complexity.

The combination of diversity and complexity – high number of species and their genotypes functioning in a web of symbiotic interrelations – results in a stable farming (Werf 1993). Some of the examples of how the mountain farmers create complexity in their farming system are as follows (after Singh and Tulachan 2001):

Cereals

1. Flour of the grains is consumed by humans.
2. Its bran is provided to milch and/or work animals.
3. Residues are fed to animals.
4. waste produced by feed is converted to manure.
5. Manure is used as an input for soil fertility (nutrition for crops).

Grewia optiva (*Bhimal*) Tree

1. Maintained as an important component of agroforestry system.
2. Leaves are used as protein-rich, tannin-free fodder for livestock.
3. Tender twigs and leaves are used as a shampoo for washing hair.
4. Fruits are used as dye.
5. Elixir prepared from its twigs and leaves is used as a medicine to facilitate parturition.

6. Its twigs and small branches are rotten to extract fiber, which is used for making ropes, mats, etc.
7. Fibre-stripped twigs and branches are use as source for light and as fuel.
8. Nutrients obtained from the tree are recycled into soil along with organic manure.

Compost

1. Husk, straw, hay, and forest litter are added to compost.
2. Used as mulch in some crops.
3. Used for fertilizing soil.
4. Used to conserve moisture in the fields and confront drought or semi-drought conditions.

Prunus armeniaca (*Chulu*) Fruits

1. Fruits are used for consumption at home.
2. Kernel is used for extracting good quality oil used for cooking and other purposes.
3. Fruit pulp is used for making jam, chutney, etc.
4. Cake of the kernel is used as scrub for washing face.
5. Cake is also used as a safe and effective poison to control rodents.
6. Leaves of the tree are used as livestock fodder.
7. Twigs and branches are used as fuel for cooking food.
8. The fruit tree is a compatible alley of the agroforestry system.

Milk

1. Used for drinking directly or through tea, coffee.
2. Converted into curd, yogurt, butter, refined butter (ghee), cheese.
3. Used for preparing a variety of sweets.
5. Sold in the market as fresh whole milk or its by-products to earn income for family.

Soil Fertility Management

"The foundations of our present scientific development were created under an explicit assumption that knowledge is a common good that is created for the common good. Exchange (of knowledge) between scientists, which is a basic tool for accelerating the creation of knowledge, is being systematically dismantled".

–Montecinos (2002)

"As the alcohol is to human beings, so are chemical fertilizers to soil".

–Bakhtawar Singh, a farmer in a Garhwal village

Marginal farmers do not regard soil as an inert, dead material, but an ecosystem, hence a part of nature's biodiversity. They follow the natural principles of nourishing the soil, so that it, in turn, nourishes the plants. They have a reverential attitude towards soil and are always cautious about its fertility status. Living soil is a precondition for a sustainable agriculture, which the marginal farmers strictly follow. In the self-contained farming systems, all they incorporate in their soil are the nutrients produced within the system. Some of farmers' strategies to manage the fertility of the soil ecosystem are given below.

Farmers transfer biomass of numerous plants in the forest ecosystems to the cropland as forest litter, or it goes after being fed to the livestock or after being used as bedding material in the animal sheds as manure. Biodiversity in the forest ecosystems will have its bearing on the soil fertility of the cropland.

Crop residues, weeds, etc., to come from cultivated land are also fed to livestock and – as manure – are recycled into the soil on which they grow. Livestock, in the process of the digestion of plant biomass in the rumen, help accelerate the process of biodegradation in nature. If the biomass were transferred to/ recycled into the cultivated land without passing it through rumen (which, functions as a wonderful micro-ecosystem), its microbial decomposition in the soil would take much longer time; and the marginal farmers cannot wait for so long.

Farmers' practice of fallowing is also meant to maintain soil fertility by containing continuous draining of soil nutrients, giving periodic chance to the land for convalescence and helping rotting of crop roots (which cannot be fed to livestock) to convert them into manure and add to soil nutrients.

Crop rotations, sequences, ways of exploiting complimentarity and synergy amongst various plant species and genetic resources are another efficient means of maintaining soil fertility marginal farmers have evolved over millennia of trials and errors.

Pest Management

"Locally developed varieties can be superior to the commercially released high-yielding varieties".

–Sundaram Verma, an innovative farmer

Marginal farmers have developed a unique mix of crop species and their genotypes that have capability of confronting the menace of pest epidemics. Very high level of inter- and intra-species biodiversity, as a law of nature, can absorb/ repel physical and biological shocks and, thus, has high degree of resilience. Further, they have selected the landraces with natural in-built resistance mechanism.

Storage bins (the *kothar*) found in the houses of marginal farmers are often impervious to moisture and insect attack. Farmers keep walnut (*Juglans regia*) and *denkan* (*Melia azedarach*) leaves along with the seeds, which serve as natural insect repellent. Pulse seeds are treated with mustard oil before storage. This helps repel insects and preserve seed for long duration. Cow urine is also used as an insecticide.

Trees farmers grow in their agroforestry systems serve as nesting and resting places for the birds, the natural predators of insects and rodents.

Weed Management

"Weeds are our friends. We include them in our foods, in the feed of our animals. Weeds have right to exist. They have existence right. We love weeds".

–Farmers of Save Seed Movement in Garhwal, India

Most of the weeds growing in the farmlands are no menace for the marginal farmers. Farmers rather view them as an indispensable component of the plant diversity. Growing of weeds is beneficial to them since these serve as animal feed and few of them as human food. Weeds also prevent soil from heating up or dying out excessively. They also induce a positive competition, which stimulates crop growth and reduces soil erosion. Such ecological and socio-economic values of weeds are well recognised by the marginal farmers in the mountains.

Jakhia (*Cleome viscosa*), a highly useful plant, grows wild, but maintained as an essential complement of the finger millet-based cropping system. From a professional agriculturist's point of view, *jakhia* is a weed since it is not cultivated. From a marginal farmer's point of view, however, it is a very useful plant and its grains are found in every household.

Some of the weeds colonise on degraded slopes. They conserve soil and, in long run, allow native successional species to emerge on the site. Ecological role of such weeds is not appreciated. However, marginal farmers give value to such weeds also.

Marginal farmers manage crop diversity in such a way that weeds would hardly compete with the cultivars. When farmers find that weeds are really competing with their main crop, they remove them simply using a hand-hoe or bullock-driven weeders. They would, in most of the cases, use this as feed for their livestock. They also look for some 'weeds' for use as vegetables; *lengura* (*Diplazium* sp.), a fern, being a common example during summer season. The unusable weeds, hardly any, would be left in the field to be rotten and mixed with the soil.

However, in recent years, invasion of exotic plants, such as *Lantana*, *Eupatorium*, *Parthenium*, and *Oxalis*, has come up as a big menace for farmers in many areas of the mountains. These exotics are replacing the indigenous plants and destroying natural biodiversity and creating an ecological imbalance in many areas, worrying the farmers. Marginal farmers are yet to cope up with this sort of menace.

References

Bahuguna, Sunderlal. 1987-88. Plant Wisdom to Restore our Eroded Earth. *Future*, 1987-88.

Bahuguna, Sunderlal. 1995. Chipko: The People's Movement with a Hope for the survival of Humankind. In Singh, Vir.; Sharma,

R.J.; and Kumar, A. (eds) *Ecological Carnage in the Himalaya*, 85-96. Dehradun: IBD.

Durhan, Jimmie. 1979. Eloheh or the Council of the Universe. *International Foundation for Development Alternative Dossier*, No. 6, April 1979.

Jackson, M.G. 2002. In *Asian Agri-History*, 6(2), p.102.

Jodha, N. S.; Banskota, M. and Partap, T. 1992. *Sustainable Mountain Agriculture.* New Delhi: Oxford and IBH Co.

Montecinos, Camila. 2002. The Ecology of Action. *Seedling*, July 2002, 11-14.

Partap, Tej. 1999. *Sustaining Livelihoods on Marginal Farms in the Hindu Kush-Himalayan Region: Challenges and Opportunities*. MFS Discussion Paper No. 1/99. Kathmandu: ICIMOD.

Powell, M. and Ikpe, F. 1992. Fertilizer Factories : Nutrient Recycling through Livestock. *ILEIA Newsletter*, 8 (3) : 13-14.

Rahamani, A.T. 2001. Is there an Imminent Crisis in Agriculture? In *An Exchange of Experieneces from South and South East Asia: Proceedings of the International Symposium on Participatory Plant Breeding and Participatory Plant Genetic Resource Enhancement*. Cali, Columbia: PRGA, IDRC, DFID, DDS, LI-BIRD, IPGRI, ICARDA.

Sardar, Ziauddin (ed). 1988. *The Revenge of Athena: Science, Exploitation and the Third World*. New York: Mansell Publishing.

Shiva, Vandana. 1989. *Staying Alive: Women, Ecology and Survival in India*. New Delhi: Kali for Women.

Singh, Vir and Tulachan, P.M. 2002a. A Dynamic Scenario of Livestock and Dairy Production in Uttaranchal Hills. *ENVIS BULLETIN: Himalayan Ecology and Development*, 10(1): 6-10.

Singh, Vir and Tulachan, P.M. 2002b. Marginal Farming in Mountain Areas. *Asian Agri-History*, 6(3): 269-280.

Singh, Vir. 1991. *Ecological Impact of Apple Cultivation in the Himalaya*. Dehradun: IBD.

Singh, Vir. 1998a. *Draught Animal Power in Mountain Agriculture: A Study of Perspectives and Issues in Central Himalayas, India*. Mountain Farming Systems Discussion Paper No. 98/1. Kathmandu: ICIMOD.

Singh, Vir. 2000. *Smallholder Dairy in Mixed Crop-Livestock Farming Systems in the Uttar Pradesh Himalayas, India*. Kathmandu: ICIMOD.

Singh, Vir. 2001. Biodiversity: A Key Issue for Commons' Management in the Himalayan Region. In Marothia, D. (ed.) *Institutionalising Common Pool Resources*, 519-530. New Delhi: Concept.

Singh, Vir. 2005. *Agrobiodiversity, Sustainability and Food Security in the Himalayan Mountains: An Uttaranchal Perspective*. Gorakhpur: GAG.

Singh, Vir. and Jardhari, Vijay. 2000. Landrace Renaissance in The Mountains: Experiences of the *Beej Bachao Andolan* in the Garhwal Himalayan Region, India. Paper presented at the *International Symposium on Participatory Plant Breeding in Asia: Exchange of Experiences*, held in Pokhara, Nepal, May 1-5, 2000.

Singh, Vir. and Tulachan, P.M. 2001. Marginal Farms and Farmers in the Himalayan Mountains: Issues and Options in Uttaranchal, India. Paper presented at the Meeting on *Issues and Options for Marginal Farmers in the Hindu Kush-Himalayan Region* organized by International Centre for Integrated Mountain Development (ICIMOD), Kathmandu.

Singh, Vir., 1995. 'Biodiversity and Farmers in Mountain Agriculture: Experiences from Garhwal Himalaya'. Paper presented at the Beijer Seminar held in Kota Kinabalu, Sabah, Malaysia from May 15-19, 1995.

Singh, Vir., 1998b. 'Traditional Biodiversity Management Strategies in the Mountains: Farmers' Experimentation in Garhwal Himalaya.' In Singh, Vir. and Sharma, M.L. (eds) *Mountain Ecosystems: A Scenario of Unsustainability*. New Delhi: Indus, 25-66.

Skolimowski, Henryk. 1995. *Dancing Shiva in the Ecological Age: Heralding the Dawn of the Ecological Reconstruction in the 21st Century*. New Delhi: Dr. Henryk Skolimowski International Centre for Eco-Philosophy/Indian Institute of Ecology and Environment.

Tulachan, P.M. 1998. *Livestock Development in Mixed Crop Farming Systems: Lessons and Research Priorities*. Issues in Mountain Development 98/5. Kathmandu: ICIMOD.

Tulachan, P.M.; and Neupane, A. 1999. *Livestock in Mixed Farming Systems of the Hindu Kush-Himalayas: Trends and Sustainability.* Kathmandu: ICIMOD and FAO.

Werf, E. van der. 1993. Ecological Farming Principles. In Singh, Vir. (ed.) *Eco-crisis in the Himalaya: Causes, Consequences and Way Out.* Dehradun: IBD.

Chapter 6

Traditional Knowledge Systems (TKS) of Water Resource Management in the Hindu Kush-Himalayan Region

A Review of India, Nepal and Bangladesh

☆ *Subodh Sharma and Otto Moog*

ABSTRACT

Traditional Knowledge Systems (TKS) in the Hindu Kush–Himalayan (HKH) region is the wisdom possessed by local inhabitants that has emerged as a means to survive under remote, isolated and harsh climatic conditions prevalent in the Himalayan region. HKH region covers some 3,500 km length over eight countries from Afghanistan in the west to Myanmar in the east. In this paper, TKS's related to water management with special focus from Bangladesh, Bhutan, Nepal, and India are described focusing on the use of rivers as a part of the European Commission supported project acronym ASSESS-HKH. TKS's reviewed are documented under the headings water harvesting methods, water abstraction methods, waterways for transportation, river restoration and water quality management methods, and water mill technology.

Introduction

Indigenous knowledge is the inter-generational wisdom of local inhabitants. The term 'indigenous' not only stands for ingrained intrinsic knowledge, but is also amenable to modifications based on latest know-how by local inhabitants through native means to suit their daily requirements. Unfortunately, such an immense treasure has been forgotten and not recognized on connotations of superstition. Recently the importance of documenting Traditional Knowledge Systems (TKS) is recognised. Bose *et al.* (1998) identified 52 indigenous practices in Bangladesh. They have compiled TKS on water harvesting, drinking water, hill irrigation systems, soil and water conservation, river transportation and forest harvesting from Bangladesh. Verma (1998) identified many practices relating to aspects like soil and water management techniques, forestry and agroforestry, livestock and fodder management etc. in north-west Himalayas of India. He has given glimpse of indigenous technology for watershed management in upper north-west Himalayas of India. Indigenous soil and water conservation practices in a participatory watershed development programmes are to be found in Premkumar (1994). 'Honey Bee' stands for people to people networking in local language devoted to indigenous knowledge systems in India. FAO (1998) describes peoples' movements in watershed development as 'Ripples of the Society' citing several case studies from different regions in India. Goonasekere and Gamage (1999) compiled indigenous practices from Sri Lanka. Kandel and Wagley (1999) compiled some salient features of indigenous technology practices for watershed management in Nepal. Indigenous technology from Bhutan are described in Honey Bee (1996) and in BIK (Bhutan Indigenous Knowledge) Newsletter published by National Environment Commission Secretariat. A glimpse on local technology knowledge for comprehensive watershed management in China is compiled by Deyi *et al.* (1998), who besides others described TKS for water harvesting and soil and water conservation techniques. Kandel and Wagley (1997) described a wide range of TKS related to natural water resources management from Nepal grouped under indigenous farmers managed irrigation systems, drinking water sources management systems, diversion channels, and indigenous milling and grinding techniques. About 80 per cent existing irrigation systems in Nepal have been initiated and developed by farmers themselves (Pant and Lohani 1983). Studies show that the area covered by farmers managed irrigation systems is more than twice as large as government managed irrigation systems in Nepal (Ansari and Pradhan 1991, Shukla *et al.* 1993, WIIAD 1993). About 90 per cent of irrigation systems in hills have been initiated and managed by farmers (WECS 1981). Indigenous grinding/milling techniques are widespread in the Hills of Nepal (Schnitzer and Melzer (1983), GATE 1988, Jantzen 1989), Aitken *et al.* 1991). Shukla and Khanal (1997) described indigenous knowledge in irrigation management: some lessons from farmer managed irrigation system in Nepal. Shrestha (1998) compiled peoples' Indigenous Technology Knowledge (ITK) for watershed management in Nepal. This thesis includes besides others ITK on water conservation, irrigation canal management, and fishing. Gurung (2002) documented ITK on water quality management from a midhill watershed in Nepal. Sharma and Gurung (2005) documented ten indigenous water quality management practices which were collected in the month of May 2002 from Galaundu-Pokhare Khola subwatershed in Dhading district, Nepal.

Water Harvesting Methods

Bangladesh

Bose *et al.* (1998) described no fewer than 52 indigenous technologies and practices for watershed management in Bangladesh. Most of these TKS's have been used for centuries by the tribal populations residing in upland areas. In water harvesting alone three ITKs are documented as indigenous water conservation, godha for water conservation, storage and harvesting in the dry season. In the hill districts of Rangamati, Khagrachari and Bandarban, water is conserved in an embankment with earthern dyke (often 5m wide and 2.5-3.5 m deep) (Figure 6.1). The water conserved in this way is used during dry period for irrigation downstream. Sometimes the whole of the river bed is converted into rice field if water availability is maintained efficiently (source: Field visit Bansa river, way to Noklab). *Godha* for water conservation (Figure 6.2) is created across a stream with bamboo and wood being used as materials for damming (about 4 m wide and 3.5 m depth) the stream.

Figure 6.1: Embankment with Earthen Dyke in Bangladesh (*Illustration Source*: Bose *et al.*, 1998)

Figure 6.2: Godha for Water Conservation in Bangladesh (*Illustration Source*: Bose *et al.*, 1998)

Nepal

The art of rainwater collection as a principal source of water has been in practice in Nepal since ancient times. This practice was most common among Brahmin community. Planting herbal plant- *Tulsi* in specially built structure called *Maeri* was considered an essential religious practice in every house and a small pond would always be built close to this for storing water. The soil excavated while constructing such pond would be used in building *Maeri* and every morning all the members of the family would use this place for bathing. The water wasted during bathing was stored in this pond and used for gardening a kitchen garden close by or else used for cattle watering. Evidence of traditional water collection ponds especially in the hilly region dates back to the historic period when Nepal was divided into several kingdoms. Many water collection ponds were built during that era. Where more flat plains were possible, bigger ponds were built with higher storage capacity and in steep places smaller ponds were built and during dry season were used as a place to perform community religious activities. Water stored in the ponds during the rainy season was used in dry periods for many domestic needs including washing, bathing, and drinking. Using roof catchments mostly made up off slates and storing run off from the hillsides for both domestic and agricultural purposes has been a practice, which came later in different parts of the country.

The technology of rainwater collection from rooftops in ferro-cement made jars for domestic purposes is becoming more and more popular in the water scarce districts of Nepal, after its successful project in Daugha Village Development Committee (VDC) of Gulmi district- introduced as a pilot project by the Rural Water Supply and Sanitation Support Programme (RWSSSP), a project jointly implemented by His Majesty's Government of Nepal and the Government of Finland (ENPHO, 1996). Then, the project has replicated this technology to many other parts like Baletaksaar and Thanapati in Gulmi district and Chahara, and Madanpokhara in Palpa district of Nepal.

Shrestha (1998) documented a system of water conservation for growing vegetables during winter in Nepal. Water from dampy land is harvested, stored and used in irrigation (Figure 6.3). Utilizing rainwater is also becoming popular in Nepal, where people can effort to bring metal sheet to collect and divert the water into a tank

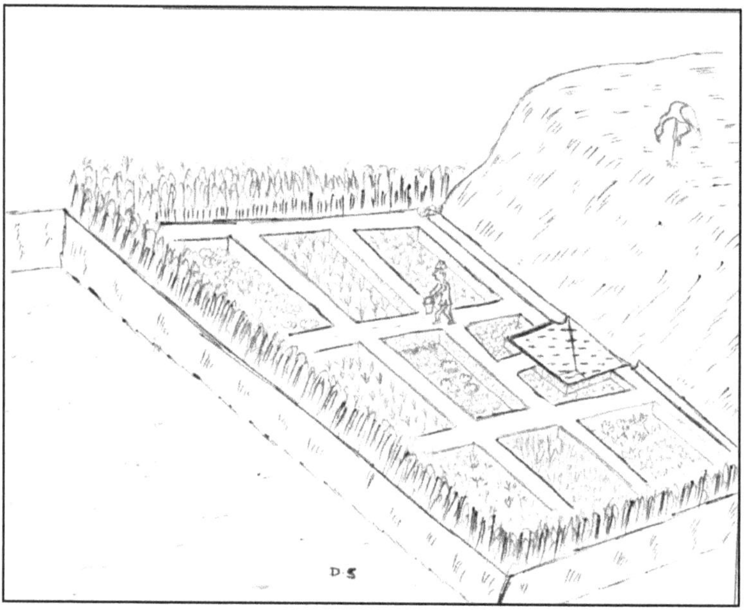

Figure 6.3: Water Conservation for Vegetables Farming in Nepal (*Illustration Source*: Shrestha, 1998)

through a gutter (Figure 6.4). Utilizing spring water for irrigation is a common practice in western part of Nepal, where a pond is dug about 6X4X2m dimension and the water that is wasted from a continuously flowing spring is collected (Figure 6.5).

India

WATMANET (1997) mentioned small reservoirs higher up in the Himalaya's, in the district of Pauri-Garhwal and the region of Ufrainkhel at an elevation of 1930. Locally called as '*Chaals*' (small reservoirs), these are commonly built in a square (about 50m X 50 m) shape and the bunds (earthen embankment) strengthened with the planting of native grasses and indigenous species of trees like the Oak, Walnut, and Rhododendron. Small ponds for spring water collection are very common in India. Water from such ponds is used for irrigating crops and also for drinking purposes. Roof water harvesting is practiced in the lower areas of Himachal Pradesh during the rainy season. Roof water is collected into a '*diggi*' (underground pit) in Kangra district also called '*Khatri*' in Hamirpur

Figure 6.4: Rainwater Harvesting in Nepal
(*Illustration Source*: Shrestha, 1998)

and Bilaspur districts (WATMANET 1997). Rainwater harvesting directly in the farm ponds or any kinds of depression is also common. The stored water is used for irrigation during droughts. In some places it is also used for human and livestock consumption. The 'changar' by which some parts of Kangra district in Himachal Pradesh are known have specific geology with sandstone rocks. The sandstone bed is considered good for harvesting drinking water. The 'changar' areas have many springs. Once the spring is identified, people clear the area and dig slightly into it to from a depression.

Water Abstraction Methods

Bangladesh

Construction of cannels for protecting the crop from inundation is a common practice among Chakma and Marma tribes in Wagga, Kaptai district of Bangladesh (Bose *et al.* 1998). A drain is constructed

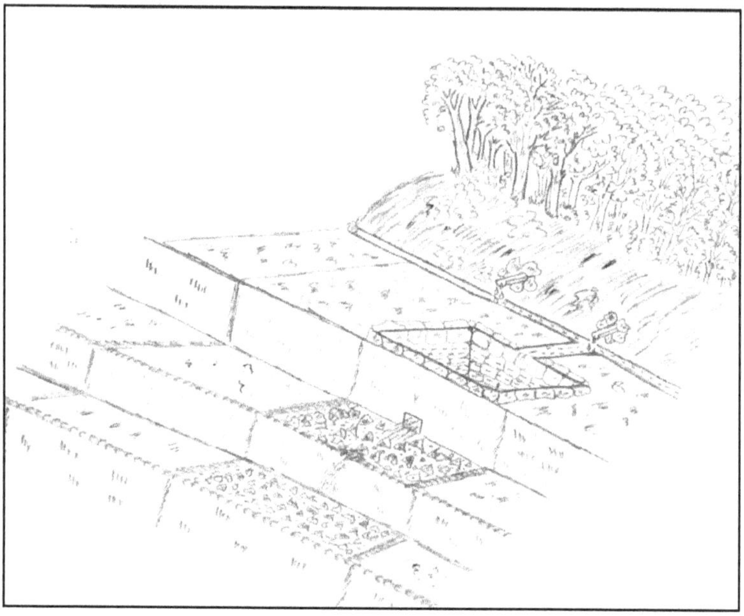

Figure 6.5: Spring Water Conservation for Orrigation In Nepal (*Illustration Source*: Shrestha, 1998)

with earth around the upper edge (the side having higher elevation) of the paddy field having high (usually 30-50 cm or more) earth wall to prevent run-off water to enter the land (Figure 6.7). Another structure used in Chemi of Bandarban is damming the rivers completely using bamboo and wooden pieces and using the excess water over the dam for irrigation purposes in the lower terraces (Figure 6.8). An indigenous structural measure is widely practiced in the terraces on slopes (locally called 'Chikon Thok') in Wagga, Kaptai district. A cluster of 6-10 bamboos are pegged by driving in 1/3 portion above normal earthern dyke.

Chikon Thok acts as strong reinforce to the dyke in case of heavy rains (Figure 6.9). Tribes in Bangladesh also use large boulders to reduce the velocity of river falling from considerable height as waterfall. These structures are about 1.6 m X 1.6 m in dimension. Stakes of bamboo or woods are also used to hold back the structures. This prevents gully formation from scouring cause by river water (Figure 6.10).

Figure 6.6: Khati or Diggi System of Water Storage in India (*Illustration Source*: Verma, 1998)

Nepal

Parallel water diversion from the same stream at different elevation along the slope is popular method of water delivery in the hills of Nepal. Farmers in Nepal design and construct channel head diversion for water abstraction using simple and indigenous

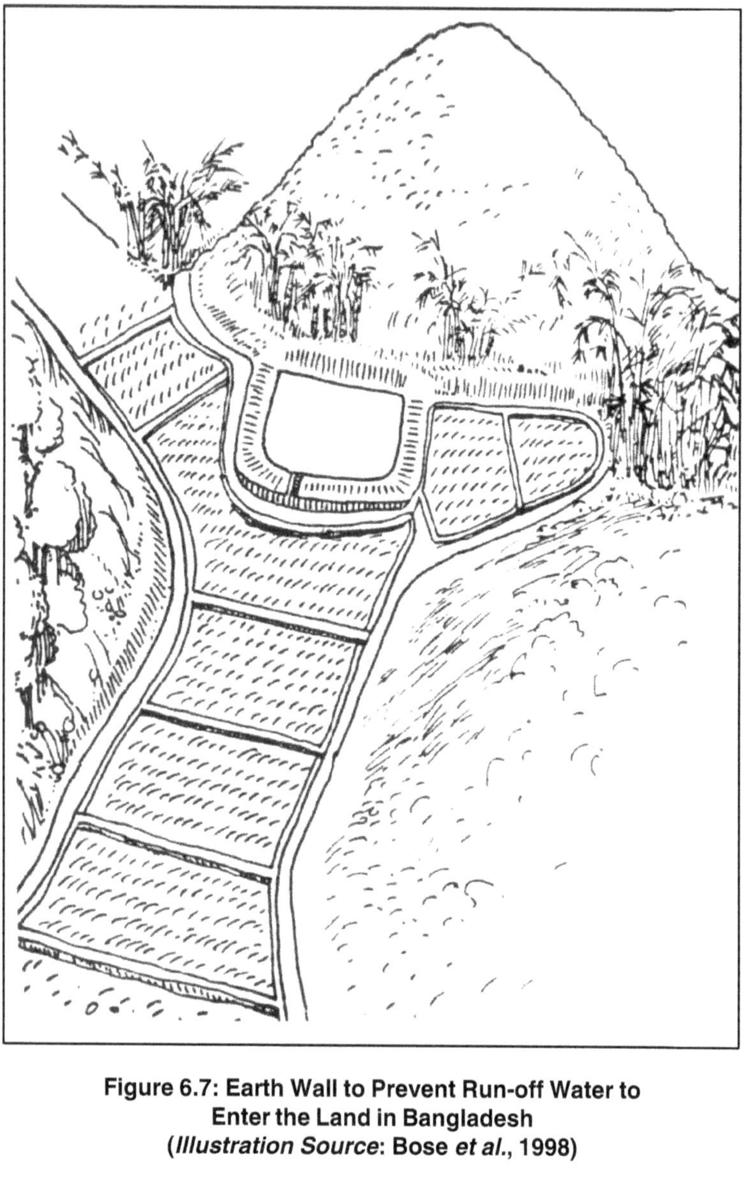

Figure 6.7: Earth Wall to Prevent Run-off Water to Enter the Land in Bangladesh (*Illustration Source*: Bose *et al.*, 1998)

Figure 6.8: Water Abstraction for Irrigation in Bangladesh
(*Illustration Source*: Bose *et al.*, 1998)

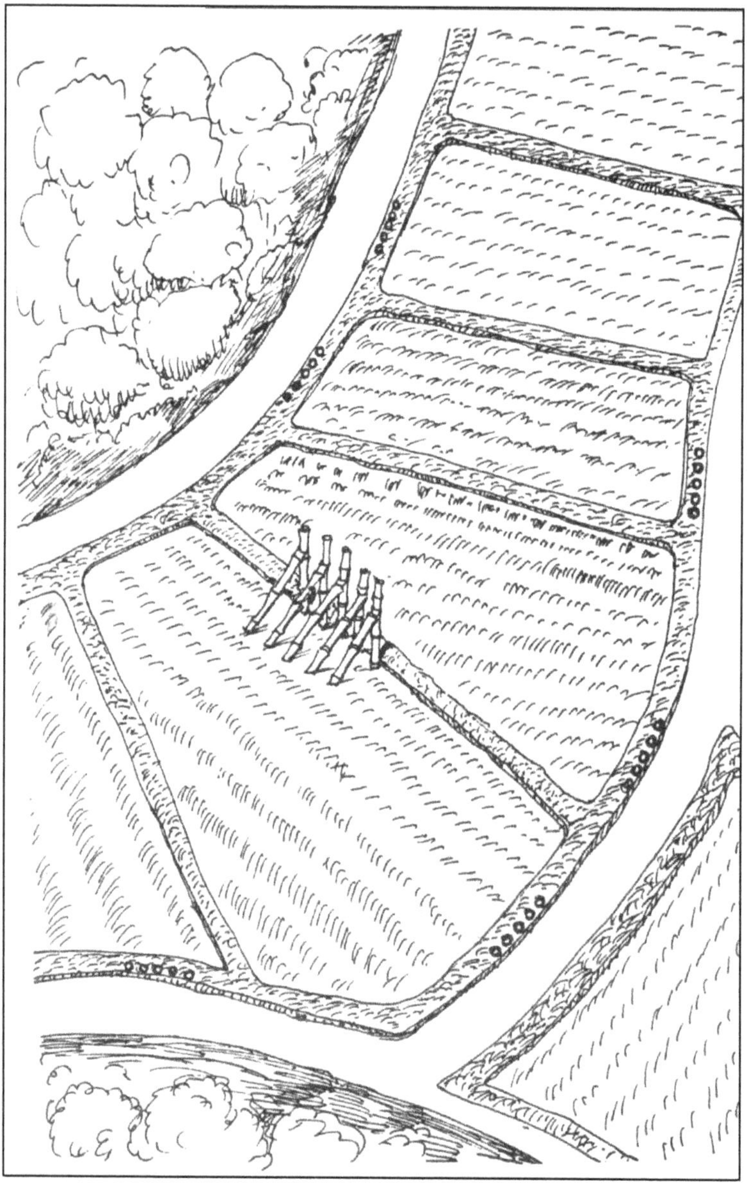

Figure 6.9: Chikon Thok Acts as Strong Reinforce to the Dyke in Case of Heavy Rains in Bangladesh (Illustration Source: Bose *et al.*, 1998)

Figure 6.10: Tribes in Bangladesh Use Boulders to Reduce the Velocity of Waterfall (*Illustration Source*: Bose *et al.*, 1998)

technologies and locally available materials preferably bushes and bamboos (Figure 6.11). In most cases the intake is found located in a narrow section that makes diversion weir reasonably stable and

Figure 6.11: Head Diversion for Water Abstraction in Nepal (*Illustration Source*: Kandel and Wagle, 1999)

cost effective (Kandel and Wagley, 1999). Where overhanging rocks are met, it is necessary to dig a tunnel, *Agris* are the group of people from Baglung and Dhading districts specialized in tunnel making. Sharma and Gurung (2005) documented a system of water delivery through bamboo made bridge, where there are gullies, cliffs or landslides enroute canals (Figure 6.12). People in the midhills of Nepal have to go long distance for collecting water for consumption. However, where sources of water are available, a bamboo split and a drain locally called 'kulo' is used to carry water by gravity as indigenous method of water abstraction.

India

Verma (1998) documented different types of indigenous techniques for irrigation water distribution for India. 'Kuhl'–open channel irrigation system are popularly known from Himachal Pradesh province, which are similar to the one described above for Nepal (Figure 6.11). To draw water in these canals stream water is diverted placing big boulders across the stream width and the water

Figure 6.12: A System of Water Delivery through Bamboo made Bridge in Nepal (*Illustration Source*: Sharma and Gurung, 2005)

leakage is sealed with mud and grass. Sometimes thorny bushes or branches of trees are put beneath the boulders, which trap sealing material flowing with the water and also provide shear strength to the diversion structure. In case the crop fields are above the stream water level, the streambed is raised (up to 5 m high) by putting big boulders in the foundation and large stone all across the rim of the diversion structure. Sand and other material brought about by the stream gradually gets deposited upstream and the streambed is suitably raised to divert the water flow to the adjoining canal. Use of strong wooden poles by fixing them vertically and horizontally across the stream width to provide shear strength to the diversion structure is also known as a rare practice of water management.

Pond irrigation is a common method to store water discharge from minor water sources for irrigation in India. Ponds are constructed on small seasonal/perennial streams involving whole stake holder community for construction of water ponding structures and mud-lined canals, and their repair and maintenance. Everyone abides by the rules of operation fixed by the community. People bring necessary equipments and a mud-and-stone structure is built across the stream to block the water flow. Such blockades are usually 5-15 m wide and 1-2 m high. Big boulders are placed at the foundation followed by smaller ones and grass and mud on the top. A hole with blocking arrangement is left at the bottom of the pond and kept close for water impounding. When the reservoir is filled the hole left at the bottom is opened and water is discharged forcefully in the canal for irrigation. The next community member now blocks the hole and uses the pond water for irrigation subsequently. One such pond is sufficient to irrigate 0.1–0.25 ha crop fields and command area of such ponds is generally 1-2 ha (Negi and Kandapal, 2002). Sometimes several smaller ponds are created along a stream to irrigate< 1 ha area and may be shared by a few community members.

Waterways for Transportation of Forests Products

Bangladesh

Transportation of harvested products using rivers as skid trails is a major disturbing factor of rivers in Bangladesh (Figure 6.13). Bundles of bamboo or thatched materials are transported as raft. Hundreds of rafts are tied together to form a type of fleet. Mainly tribal use this mode of transportation for carrying their products to the market places. However, Bose *et al.* (1998) reports this method of transportation in Karnafully paper Mill at Chandroghona, Rangamti, Khagrachari, and Bandarban. Similarly, in the eastern part of the country, the tribal people use indigenously made boats or dugouts canoes extensively. The dimension of such boats varies from 5.5 to 20 m in length (Bose *et al.*, 1998).

River Restoration and Water Quality Management Methods

Bangladesh

Tribes in Bangladesh, especially from Bandarban, have restrictions to cut trees along the channel or a watercourse and maintain a strip of vegetation at some distance (Figure 6.14). These restrictions in felling help to maintain water quality in the streams and channels and also control soil erosion (Bose *et al.*, 1998).

Figure 6.13: Transportation of Harvested Products Using Rivers as Skid Trails in Bangladesh (*Illustration Source*: Bose *et al.*, 1998)

Figure 6.14: Tribes in Bangladesh, have Restrictions to Cut Trees along the Watercourse (*Illustration Source*: Bose *et al.*, 1998)

Nepal

Use of loose boulders for reducing the cutting effect of streams locally called as 'bhakari' is used by farmers to protect their paddy fields in the hills of Nepal and India. Four pillars are used at four corners, which are mostly made of wood. Alternatively trees of *Salix* species are grown abundantly along the river bank, which act as buffer to improve the quality of river.

India

Use of loose boulders spurs for reducing soil erosion (Figure 6.15) is also common in India (Verma, 1998). In addition use of loose boulders diversion dam with spillway in centre is also very common

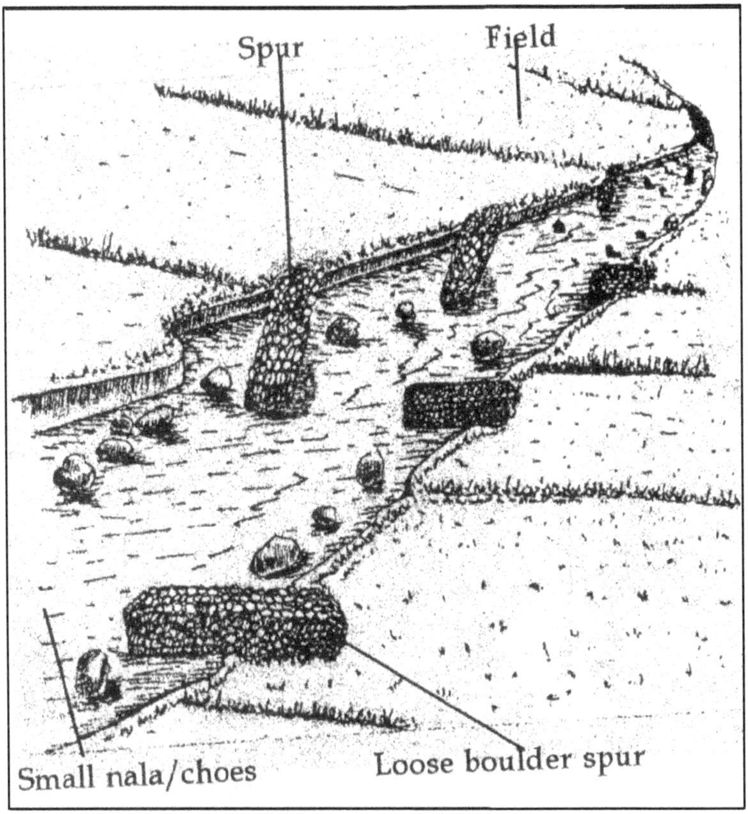

Figure 6.15: Use of Loose Boulders Spurs for Reducing Soil Erosion is Common in India (*Illustration Source*: Verma, 1998)

in lower areas of Himachal Pradesh (Figure 6.16). These dams are constructed across the streams for controlling soil loss from stream bed.

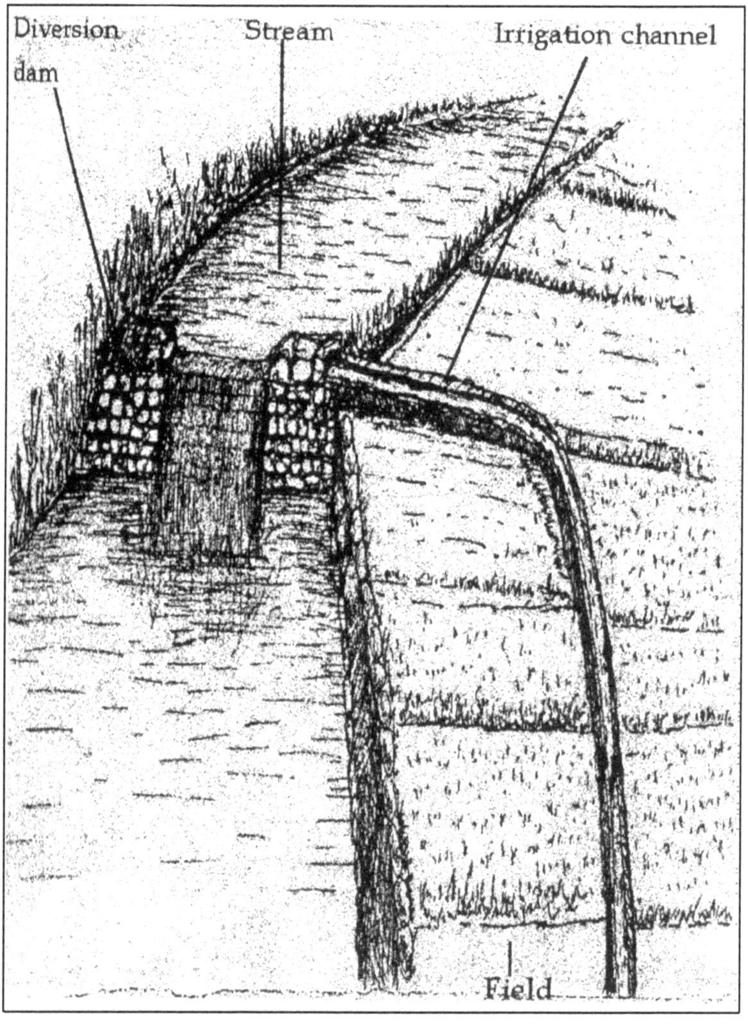

Figure 6.16: Use of Loose Boulders Diversion Dam with Spillway in Centre is also Found in India (*Illustration Source*: Verma, 1998)

Methods of Irrigation in HKH

Nepal

Drip irrigation is not common in Nepal. In most of the farmers managed irrigation systems in Nepal, temporary and semi-permanent type overflow weirs are found to be performing satisfactorily. Brushwood temporary weir (Figure 6.17) is the most common diversion structure built with locally available materials. The practice of irrigation by continuous flowing spring water in paddy is common in Nepal (Shukla and Khanal, 1997). In the mountains this practice helps in increasing the soil temperature and in lowland to lower down the soil temperature. Practice of tapping flash flood in the beginning of rainy season is also reported in Nepal. This brings fertile silt that helps in adding nutrients to soil. In many farmers managed irrigation systems in Nepal, for proportionate division of water share, device called Sancho or Jhyal, which is made of wooden blocks with rectangular notches (Yoder *et al.* 1987) is popular. Where such devices are not in use, water allocations is dependent on Kulara meaning delivery through each branch canal ensured by the size of inlet opening.

'Palo bhadhney' is also a common method of water allocation, where whole community is divided on the basis of farm families and one family gets one full water day or limited to few hours to irrigate their fields turn wise.

India

Similarly in India the management of water is regulated by dividing the field into different compartments. The mouth of first compartment is closed to regulate the flow of water towards the second compartment and so on.

In Kinnaur channels passing through a village are harvested on turn basis called 'pala' as described earlier for Nepal. In the community run canals the community elders fix schedule of irrigation depending upon the crops, season and availability of water. Priority is given to vegetable crops for irrigation. If the water discharge is low, irrigation is done from tail to head of the canal. Each field owner waits for his turn for irrigation. It may come any time during day or night. When plenty of water is available irrigation depends upon the choice of the farm owner and there is no fixed rotation. In

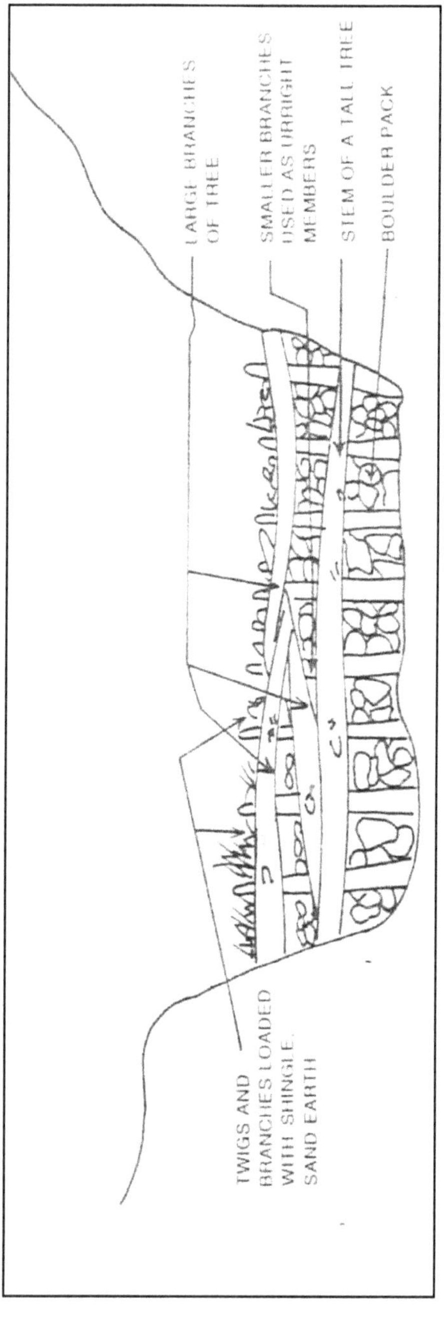

Figure 6.17: Brushwood Temporary Weir is Com mon in Nepal (Kandel and Wagle, 1999)

this situation water is diverted for irrigation at many points across the canal. Sometimes, when the flow is very low the main canal is blocked and water is allowed to fill in through the canal length. Once it is filled up, the blockade is removed suddenly to use the force of water to travel the desired canal distance rapidly. Sometime more than one peasant shares the water thus stored at one time.

Farmers in India irrigate crops based on traditional knowledge. In paddy, irrigation is done 7-8 times throughout the growing period when normal water supply is available. Irrigation is preferred either early morning or late evening to escape the warming up of irrigation water due to sunshine, which is considered harmful for the crop. Sometimes when plenty of water is available, small stream of water flowing is left for most oft he growing period (July-August) in the rice fields and an outlet from the corner of the fields for the overflow water is left. This continuous supply of fresh water is considered even more beneficial for paddy. For wheat, irrigation is given at 20-25 days interval and altogether 5 irrigations are preferred for the wheat crop. Light irrigation during winter is allowed and water flow is guided to spread over the field uniformly to avoid excessive wetting, which makes the soil cold, and retards tillering and growth of wheat. For tuberous crops such as, potato, onion, radish, turmeric, ginger etc. irrigation at 20 days interval during winter and at 7-8 days interval during summer is practiced.

Watermill Technology

Bhutan

In Bhutan, as large population are Budhists, use of small springs in running payer wheels is common. Water from springs is diverted using the natural gravitational flow of water to run such wheels, which resemble 'ghatta' in Nepal and 'gharat in India (Figures 6.18 and 6.19).

Nepal

Traditional watermill (*Ghatta*) has been use in Nepal for grinding of wheat, maize, millet etc. Stream is diverted into channels with piled up stones and brushwood weirs. Last part of the channel leads into a chute at an angle of about 45degree. Usually there are six to fourteen blades in a turbine (Kandel and Wagle 1999). WATMANET

Figure 6.18: A Typical House where Traditional Water Mills are kept (Verma, 1998)

(1997) has elaborated the methods, advantages and disadvantages of indigenous water mills in Nepal.

India: Traditional watermills in India are locally called as *'gharat'*. Diversion on streams called *'kuhl'* is also common in India for running such mills. Thesedays *'gharats'* are becoming rarer and are in an extinct stage, because of power supply availability (Verma, 1998).

Recommendations

The traditional knowledge (TKS) and skills of people is diminishing at an alarming rate in the Hindu Kush- Himalaya. In order to protect this knowledge it should be made available to the less knowledgeable within the community. Individuals can use research, literature, and even own experience to enhance TKS and can be used for commerce or the advancement of science. Traditional knowledge should be used to offer solutions to common problems within the community and not that it is forgotten in the name of superstitions or unscientific judgments.

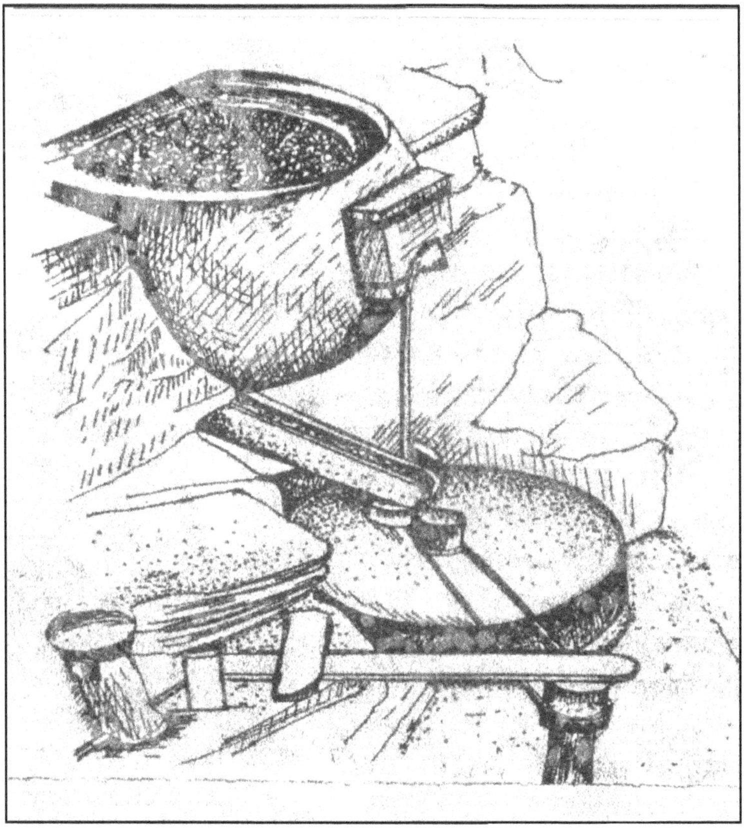

Figure 6.19: Structure of a Water Mill (Verma, 1998)

References

Aitken, J. M. Cromwell, G. and Wishart, G. 1991. Mini and Micro-Hydropower in Nepal. *ICIMOD Occasional Paper* No. 16, December 1991, ICIMOD, Kathmandu, Nepal.

Ansari, N. and Pradhan, P. 1991. *Assistance to Farmers' Management Irrigation Systems: Experiences from Nepal.* Ministry of water resources, DOI, Nepal.

Bose, S. K.; Ghani, O.; Emdad Hossain A. T. M.; Mridha N. N. and Muhammad T. (1998): A compilation of indigenous technology knowledge for upland watershed management in Bangladesh.

PWMTA-FARM Field Document No. 11. Published by Participatory Watershed management Training in Asia (PWMTA) program GCP/RAS/161/NET, FAO (UN), U. N. Bldg. P. O. Box 25, Kathmandu, Nepal.

Deyi W., Xiaoying, L., Sobei, S., Zifieng, B., Rongen, L., and Yuhua, W. 1998. *A Glimpse of Local Technology Knowledge for Comprehensive Watershed Management in China*. Edited by P. N. Sharma, IRTCES (International Research and Training Center on Erosion and Sedimentation, Beijing, China.

ENPHO 1996. A survey report on water quality testing from different rainwater collection systems and other water sources in Daugh VDC, Gulmi, Nepal.

FAO 1998. Ripples of Society. PWMTA and FARM programme. Field Document No. 14. Gandhi Peace Foundation, New Delhi, India.

Ferroukhi, L. and Chokkakula, S. 1996. Indigenous knowledge of water management. Reaching the unreached: Challenges for the 21st century. 22nd WEDC Conference, New Delhi.

GATE 1988. Micro Hydropower-Questions, Answers, Information. No. 1/88. March 1988, Eschborn, Federal Republic of Germany.

Goonasekere, K. and Gamage, H. (1999): Some indigenous technology knowledge and practices for watershed management in Sri Lanka. *PWMTA Field Document* No. 18.

Gurung, B. 2002. Documentation of Indigenous Technical Knowledge for water quality assessment in Jhikhu Khola watershed. Submitted to Dr. Subodh Sharma, Kathmandu University.

Honey Bee 1996. Indigenous technology from Bhutan. Honey Bee Vol 7(4) October-December.

Jantzen, D. and Koirala, K. 1989. *Micro-Hydropower in Nepal*. Bikash Enterprises Inc., Kathmandu.

Kandel D. D. and Wagley, M. P. (1999): Some salient indigenous technology practices for watershed management in Nepal. *PWMTA Field Document* No. 19.

Kandel, D. D. and Wagley, M. P. 1997. *Indigenous Technological Knowledge for Watershed Management in Nepal*. Department of soil conservation and watershed management, HMG/Nepal.

Negi, G. C. S. and Kandpal K. D. 2002. Traditional methods of water management inthe central Himalayan agriculture. *Indian Journal of Traditional Knowledge* Vol. 2(1), July 2003, pp. 256-264 G. B. Pant Institute of Himalayan Environment and Development, Kosi-Katarmal, Almora 263 645 (Uttarakhand), India

Nepal Agroforestry Foundation (NAF) (1999): Galaundu/Pokhare Khola sub-watershed community resource management plan. Dhading, Nepal.

Pant, T. N. and Lohani, J. 1983. Some observations on irrigation development in Nepal. Paper presented at the seminar on water management isues. 31 July to 2 August 1983. Jointly organized by Ministry of Agriculture, HMGN/APROSC and ADC.

Premkumar P.D. 1994. Indigenous soil and water conservation practices in a participatory watershed development programme. In *Farmers are Engineers*, Edited by Blaise Humbert-Droz, Swiss Development Corporation, Banglore-560 001, India.

Schnitzer, V. and Melzer, A. 1983. Activating traditional knowledge techniques: Development and improvement of the Nepalese water mills "Ghatta". GTZ, Kathmandu.

Sharma, S. and Gurung, B. 2005. Indigenous water quality management practices in Dhading district, Nepal. *Journal of Environment* Vol. 9, No. 10, Ministry of Environment, Science and Technology, HMG/N, Kathmandu, Nepal.

Shrestha, S. 1998. A compilation of Peoples' Indigenous Knowledge (ITK) for watershed management in Nepal. Project report submitted in partial fulfilment of the requirements for the degree of Bachelor of Science (Hons.) Environmental Science, Department of Biology and Applied Biology, School of Science, Kathmandu University, Nepal.

Shukla, A. and Khanal, P.R. 1997. Indigenous Knowledge in Irrigation Management: Some Lessons from Farmer Managed Irrigation System. In Nepal. Paper presented at the National Workshop on "Indigenous Technological Knowledge for Watershed Management" organised by Department of Soil Conservation, HMG/N at Institute of Forestry, Pokhara on December 25-27, 1997.

Shukla, A. K., Gajuel, K.P., Shivakoti, G. P., Poudyal, R., Pandit, K. N., Adhikari, K. R., Thapa, T. B., Shakya, S. M., Yadav, D. N., Joshi, N.R., and Shrestha, A. P. 1993. *Irrigation Resource Inventory of East Chitwon*. IMSSG/IAAS, Rampur, Chitwon, Nepal.

Verma, L. R. 1998. A glimpse of indigenous technology knowledge for watershed management in upper north-west Himalayas of India. PWMTA *Field Document* No. 15. Published by Participatory Watershed management Training in Asia (PWMTA) program GCP/RAS/161/NET, FAO (UN), U. N. Bldg. P. O. Box 25, Kathmandu, Nepal.

WATMANET 1997. Asian Watmanet (Watershed Management Network) Newsletters, Issue 1-15, Dec., 1994-June, 1998. Published by Participatory Watershed Management Training in Asia (PWMTA) Program, Netherlands/FAO, GCP/RAS/161/NET, Kathmandu, Nepal.

WECS 1981. *Irrigation Sector Review.* Report no. 312/190981/1/1. Water and Energy Commission Planning Unit, HMG/Nepal.

WIIAD 1993. Policy analysis in agriculture and related resource management. Nepal agri-brief, series no. 3 and 4. *Quarterly Progress report*. January-march, 1993, Kathmandu.

Yoder R., Martin E., Barker R., and Steenhius, T. S. 1987. Variation in Irrigation Management Intensity: Farmer Managed Hill Irrigation System in Nepal. Water *Management Sybthesis Report* No. 67. USAID.

Chapter 7

Organic Farming: An Indigenous Technology for the Hindu Kush-Himalayan Agriculture

☆ *Pinki Bhandari*

With sharply increasing food grain demand and limited land resources, the agricultural growth has shown a sharp inspiring rise since last three decades. These great achievements are the direct result of new revolutionalized technology in agriculture in the way towards developments of improved germplasm of several crops, expansion of irrigation facilities, increased use of safer agro-chemicals and increased cropping intensity. Indian agriculture has attained a new height of 209 Mt (in 2003-2004) in food grain production in comparison to 1.4Mt in 1947. Despite this raising curve, a lot more scope to enhance production is there as the yield level for each crop is still has a comparable gap with productivity in various parts of the world.

Along with a boon of introduction of high yielding crop varieties a blast decrease in soil structure and quality have been observed. This has resulted into heavy reliance on chemical fertilizers. Thus, the fertilizer consumption increased from 66,000t nutrients in 1951-52 to about 20.15 Mt nutrients in 2003-2004. But the harvested crops remove about 27Mt during 2003-2004, thus leaving a deficit of 9.25Mt of N, P_2O_5 and K_2O during this corresponding year. These trends in nutrient consumption and removal along with resultant alteration in ratio between N, P_2O_5 and K_2O has further aggravated the situation of soil mining.

At national level the consumption ratio of N : P_2O_5 : K_2O is 8.5 : 3.1 : 1 against ideal 4 : 2 : 1. This imbalance in fertilizer consumption pattern in the country needs great regulatory Institutes and organizations.

Various surveys conducted at state and farmers level by ICAR, SAUs and Institute organizations have revealed a great deterioration in soil fertility with a large scale decline in organic matter. But the only use of organic manure application may not be sufficient to combat the ever increasing population in developing countries. Thus, the integrated nutrient management is the only alternative to maintain the soil productivity without influencing the total food grain production in the country.

Integrated nutrient management is the combined application of chemical fertilizers along with organic manures, green manures, bio-fertilizers and other organic recyclable materials for crop production. Being ecological, social and economically viable and environmentally healthy technology, this can be adopted by farming community to get higher productivity while maintaining soil fertility. These management practices not only encourage the use of in-house organic waste but also save on the cost of fertilizers for crop production.

In perusal of the collected data, it has been estimated that about 15Mt of N_2 + P_2O_5 + K_2O will be available as per projections on the availability of organic resources in India during 2000 out of which 2.65Mt can be exploited for recycling (Tandon, 1997).

Organic Agriculture in Himalayan region

"Organic agriculture is a holistic production management system which promotes and enhances agro-ecosystem health, including biodiversity, biological cycles and soil biological activity.

Typically, farmers experience some loss in yields after discarding synthetic inputs and converting their operations to organic production. Before restoration of full biological activity (*e.g.* growth in beneficial insect populations, nitrogen fixation from legumes), pest suppression and fertility problems are common.

Sometimes it may take years to restore the ecosystem to the point where organic production is possible. In these cases other sustainable approaches that allow judicious use of synthetic chemicals may be more suitable start-up options. One strategy involves converting farms to organic production "in installments", so that the entire operation is not put at risk. Aim is to optimize land, animal, and plant interactions, preserve natural nutrient and energy flows, and enhance biodiversity, all of which contribute to sustainable agriculture.

Use of crop rotations, organic manure and mulches improves soil structure and encourages development of a vigorous population of soil micro-organisms. Mixed and relay cropping provides a more continuous soil cover and thus a shorter period when the soil is fully exposed to the erosive power of the rain, wind and sun.

Organic farmers also employ natural pest controls–*e.g.* biological control, plants with pest control properties–rather than synthetic pesticides which, when misused, are known to kill beneficial organisms, cause pest resistance and often pollute water and land.

Crop rotations encourage a diversity of food crops, fodder and under-utilized plants which, in addition to improving overall farm production and fertility, may assist in the on-farm conservation of plant genetic resources.

Organic Manures

Organic manures are a valuable source of available nitrogen (N). However, using either too much or too little N can adversely affect crop yields and quality. By integrating manure and inorganic fertilizer N additions, crop yields and quality will be optimized and pollution risks reduced. It acts as a soil conditioner by inducing microbial activities which improves air-water relationships, increases shelf-life, fertility and resultant soil become resistant to environmental stresses such as drought, extreme heat, early frost, compaction, erosion and pest and disease problems.

The ammonium (readily available) N content of manures can easily be lost to the atmosphere by ammonia volatilization. To reduce losses:

- Manures can be rapidly incorporated into the soil
- Slurries can be applied using injection or band spreading methods

- Losses are potentially greatest from slurries and poultry manures, which contain larger amounts of readily available N than straw-based farmyard manures.
- The amount of N loss depends on:
  - Manure application rate
  - Rainfall volume after application, and
  - Soil type

Manure Details

Manure Type	Cattle, FYM
Application rate	35t/ha
Dry matter content	25 per cent
Total N	6kg/tonne
Readily available N	1.5kg/tonne

The continuous application of FYM found to eliminate the application of phosphate fertilizers as the available P content of soil had built up to as high as 21mgkg^{-1} at which the response to P is not expected (Gupta *et al.*, 1999)

Both conventional and organic agriculture utilize manure as part of regular farm soil fertilization programs. Certified organic farmers, however, must have a farm plan detailing the methods used to build soil fertility including the application of manure or composted manure. Certified organic farmers are prohibited from using raw manure for at least 60 days before harvest of crops grown for human consumption

Composted Manure

It is a primary source of soil fertility for organic farmers. It offers a natural means to cycle plant nutrients. As such, animal manure forms an important part of organic soil fertility programs

Compost

Compost is the product of a carefully managed aerobic process by which organic materials are digested by microorganisms. Organic materials added to the composting process are limited to those permitted for crop production by this (organic) rule, such as manure, plant residues, and mulch. Production inputs made by the addition of synthetic materials beyond an incidental residue, municipal solid waste that is not source separated, and sewage sludge are not considered compost for the purpose of this (organic) rule. This organic material must be managed to reach temperatures for the duration necessary to effectively stabilize nutrients and kill human pathogens. At a minimum, the process requires that organic matter reaches temperatures of at least 130 degrees F. (55°C), for a period of several days, be thoroughly mixed, and achieve that temperature again before finishing for a period of approximately six weeks.

E. coli, *Salmonella*, and other pathogens found in manure can be reduced by proper composting. Compost should be maintained at temperatures of 55-60°C (130–149°F) for a period of several days if possible up to two weeks. Composting reduces pathogens in several different ways. One way is by generating temperatures unfavorable to the undesirable organisms. Temperature increases during composting are the result of microbial metabolism. The temperatures generated have the benefit of reducing populations of many pathogenic organisms.

Crop Rotation

Crop rotation means changing the kind of crop produced in each field each growing season. Organic farmers may take five or more years before the same crop is planted again in a particular field. Organic farmers, including livestock producers, practice crop rotation in order to build and maintain soil health and to break the lifecycle of pests, thus reducing the need for synthetic fertilizer and pesticide applications.

"Green Manure" or Cover Crops

Green manure is a crop that is grown, then plowed into the soil or otherwise left to decompose for the purpose of soil improvement. Examples of cover crops used for green manure include soybeans, clover, rye, and others. Green manure does not mean raw manure.

Manure Management

Manure offers a natural means to cycle plant nutrients. As such, animal manure forms an important part of organic soil fertility programs. Manure, either by it or blended with crop residues, makes up much of the raw material for the compost used on organic farms.

Non-synthetically Treated Minerals

Organic farmers may use minerals such as limestone, rock phosphate lignite (a sulfate of potash- magnesia), green sand, rock dust and others to improve the soil's tilt. The tilt of soil is a composite of its texture, structure, aggregation, density, drainage, and water holding capacity.

Green manuring with Sesbania aculeate in conjunction with 2.5 gypsum/ha 50 per cent GR) found in maximizing rice and wheat yield and also improving soil fertility (Swarup, 1998).

Press mud is another by product from sugarmills, is an effectives ameliorative in reclamation of soil solidity. Coirpith, a by product of coir industry constrains nutrients and high lignin (N=0.26 per cent, P=0.01 per cent, K=0.78 per cent, lignin=30 per cent, cellulose=26 per cent). Compositing of coirpith is practiced with Pleurotus and urea for reducing C: N ratio and the bulk of the material.

The city sewer along with industrial effluent can also be exploited as organic manure and its applicability will depend upon the quantity of toxic metals present in them. It was observed that application of sewage not only influenced the yield of rice but also improved the soil fertility. The incorporation of crop residues in conjunction with green manure crops increased the yield of rice (Singh and Singh, 1996).

Food and Agricultural Organization (FAO) has developed some IPNS models for selected cropping systems (Ray and Ange 1991). The fertilizer recommendations for rice-wheat and maize wheat is 249kgN+90kgP_2O_5+120 kg K_2O with an expected potential grain yield under optimum management.

Vermicompost: A Novel Method for Himalayan Agriculture

The development of ecofriendly novel processes are based upon the utilization of biological systems. One such system involves the culture of earthworms (vermiculture) for the stabilization of a variety of organic solid wastes (vermicomposting). The key role of earthworms in improving soil fertility is well known since long (Darwin 1881; Chandana 1981; Kale and Krishnamoorthy, 1981).

Earthworms feed on any organic waste, consume two to five times their body weight and after using 5-10 per cent of the feedstock for their growth, excrete the mucus coated undigested matter as wormcasts. Wormcasts consist of organic matter that has undergone physical and chemical breakdown through the activity of the muscular gizzard which grinds the material to a particle size of 1-2 micron. The nutrients present in the wormcast are readily soluble in water for the uptake of plants (Bhawalkar and Bhawalkar 1993). Vermicasting is a rich source of macro and micronutrients, vitamins, enzymes, antibiotics, growth hormones and immobilized microflora.

Vermicompost refers to organic manure produced by earthworms. It is a mixture of worm castings (faecal excretions), organic material including humus, live earthworms, their cocoons and other organisms. Vermicomposting is an appropriate technique for the disposal of non-toxic solid and liquid organic wastes. It helps in cost effective and efficient recycling of animal wastes (poultry, horse, piggery excreta and cattle dung), agricultural residues and industrial wastes using low energy (Jambhekar 1992).

The main steps of vermicomposting are described below:

1. *Selection of earthworm:* Earthworm which is native to the local soil and vermicompost may be used.
2. *Size of pit:* Any convenient dimension such as 2m x 1m x 1m may be prepared. This can hold 10 -40 thousand worms giving one tonne manure per month.
3. *Preparation of vermin-bed:* A layer (15 -20 cm thick) of good loamy soil above a thin layer (5 cm) of broken bricks and sand should be made. This layer is inhabited by earthworms.
4. *Inoculation of earthworms:* About one hundred earthworms are introduced as optimum inoculating densities into a compost pit of about 2m x 1m x 1m, provided with a vermibed.
5. *Organic layering:* It is done on the vermibed with fresh cattle dung. The compost pit is then layered to about 5 cm with dry leaves or hay. Moisture content of the pit without flooding is maintained through the addition of water.
6. *Wet organic layering:* It is done after four weeks with moist/ green organic waste which can be spread over it to a thickness of 5 cm. This practice can be repeated every 3-4 days. Mixing of wastes periodically without disturbing

the vermibed ensures proper vermicomposting. Wet layering with organic wastes can be repeated till the compost pit is nearly full.

7. *Harvesting of vermicompost:* At maturation, the moisture content is brought down by stopping the addition of water for 3-4 days. This ensures drying of compost and migration or worms into the vermibed. The mature compost, a fine loose granular mass is removed out from the pit, sieved, dried and packed.
8. *Rate of application:* Mature vermicompost is recommended @5t/ha (Ismail 1994)

Nutrient Contents of Vermicompost

The nutrient status of vermicompost (prepared by two species of earthworm) and farm yard manure (FYM) is shown in he Table below for comparison.

Parameters	*Eisenia foetida*	*Perionyx excavatus*	*FYM*
pH	7.40	7.00,	7.20
Organic Carbon (per cent)	27.43	30.31	12.20
Total nitrogen (per cent)	0.60	0.66	0.55
Total phosphate (per cent)	1.34	1.93	0.75
Total potassium (per cent)	0.40	0.42	2.30
C : N ratio	45.70	45.90	24.40

Source: Shinde *et al.* (1992)

There are appreciable differences in the nutrient content of vermicomposts produced by different species of earthworms. Vermicompost is thus not a single, standard material or product. On an average, vermicompost contained more C and P than FYM, it had less K and micronutrients than FYM and both had comparable contents of N. Vermicomposts generally had wide C: N ratio as compared to FYM. On the whole vermicompost cannot be described as being nutritionally superior to other organic manures but the unique way in which it is produced, even right in the field and at low cost makes it very attractive for practical application (Gaur and Singh, 1995).

The most common method of solid waste disposal is land spreading which causes pollution of soil as well as surface and ground water resulting in mortality of aquatic organisms. Vermicomposting of wastes controls the pollution of soil and water, thus ensures the survival and growth of fish, prawn and other aquatic organisms. The application of vermicastings which is a high grade organic fertilizer, to the aquaculture ponds reduces the input cost making the aquaculture production process more profitable and it also helps in controlling the harmful effects of chemical fertilizer application. Deolalikar and Mitra (1996) have used vermicompost prepared from paper mill solid waste for fertilizing aquacultural tanks and found an increase in net primary productivity from 32.08 to 220.83 mgC/m/h. Vermicompost application also showed better growth of rohu fish (*Labeo rohita*) when compared with other commercially available organic manures (Deolalikar and Mitra, 1997). There is an increasing demand for protein-rich raw materials in fish and other animal feed industry. Fish meal is the main protein component of fish feed. Earthworm is generally used as bait by the anglers. But large scale vermiculture has the potential of supplying earthworm meal as a substitute of fish meal. The earthworm meal contains all the essential amino acids required in fish feed. The methionine and lysine availability are higher than that of fish meal.

Vermicomposting technology is applicable to the rural as well as the urban society. It not only helps in commercial aquafarming but also acts as a convenient source of earthworm for growing ornamental fishes in aquarium. Thus vermicomposting can be included as a component of sustainable life-style. Application of vermiculture and vermicomposting in aquaculture is ecofriendly and bioethically acceptable.

References

Chandana, S.S. 1981. Relationship between earthworm and soil fertility. *Farmer and Parliament*, January, 15-16.

Bhawalkar, U.S and Bhawalkar, U.V. 1993. Vermiculture biotechnology. In: *Organics in Soil Health and Crop Production* (Ed. P.K. Thampan), Peekay Tree Crops Development Foundation, Cochin, 65-85.

Darwin, C.R. 1881. The formation of vegetable mould through the action of worms with observations on their habits, Murray, London, pp 298.

Deolalikar, A.V. and Mitra, A. 1996. *Vermicompost of paper mill solid waste: a potential low cost source of organic manure for primary production in aquaculture*, Aquaculture (in press)

Deolalikar, A.V. and Mitra, A. 1997. Application of paper mill solid waste vermicompost as organic manure in rohu (*Labeo rohita*, Hamilton) culture: A comparative study with other commercial organic manure. *Aquaculture* (communicated)

Gaur, A.C. and Singh, Geeta (1995): Recycling of rural and urban wastes through conventional and vermicomposting. In Recycling of crop, animal, human and industrial waste in agriculture (Edt. H.L.S. Tandon), FDCO, New Delhi, 31-49.

Gupta,A.P., Narwal, R.P and Antil, R.S. 1999. Direct, residual and cumulative effect of farmyard manure application on the nitrogen response to pearl millet- wheat cropping sequence. pp. 137-140. In Proceedings International Symposium on long terms Field Experiments held at Martin- Luther University Halle, Germany.

Gupta, A.P and Jagannath.2000a. Integrated nutrient management for higher productivity and maintenance of soil health. Haryana Fmg. 30(3): 3-4.

Ismail, S. (1994): Vermitech: the use of local species of earthworms in agriculture. *Changing Villages*, April-June, 27-31.

Jambhekar, H.A. 1992. Use of earthworm as a potential source to decompose organic wastes, in *Proc. Nat. Seminar on Organic Farming*, MPKV, Pune, 52-53.

Kale, R.D. and Krishnamoorthy, R.V. (1981): Enrichment of soil fertility by earthworm activity, G.K.V.K, *UAS Tech.* no. 37, 64-68.

Nagaraja, D.N and Krishnamurthy, K.1989. Sewagw-cum-fertilizer on the growth and yield of rice. *Mysore J. Agric. Sci.* 23:289-291.

Reinecke, A.J. and Alberts, J.N. 1987. The chemical and amino acids composition of the compost worm *Eisenia foetida* (Oligochaeta) as potential source of protein for animal feed. *S.A. Tydskrif Vir Natuurwetenskap en Tegnologie*, 6, 1-14 (translated copy).

Roy,R.N and Ange,A.L.1991. Integrated Plant nutrition system (IPNS) and sustainable agriculture. Proc. FAI Annual Seminar, FAI, N.Delhi. pp SV/1SV/1-12.

Shinde, P.H., Naik, R.L., Nazirkar, R.B., Kadam, S.K. and Khaire, V.M. 1992. Evaluation of vermicompost. *National Seminar on Organic Farming*, MPKV, Pune, 54-55.

Tandon, H.L.S. 1997. Organic Resources: An assessment of potential supplied their contribution to Agriculture and Policy Issue for Indian Agriculture from 2000 to 2025. Plant Nutrient Needs, Supply, Efficiency and Policy Issues: 2000-2025. pp 15-28.

Tandon, H.L.S. (cd) 1995. *Recycling of Crop, Animal, Human and Industrial Wastes in Agriculture*. Fertiliser Development and Consultation Organization, New Delhi, India.

Swarup, A.1998. Emerging soil Fertility Management Issues for susyainable Crop Productivity in Irrigated System.pp 54-68. In A. Swarup, D. Damodar Reddy and R.N.Prasad (eds) Proceedings of a National Workshop on long term Soil Fertility Management through Integrated Plant Nutrient Supply.Indian Institute of Sciences, Bhopal, India.

Singh, B and Singh, Y.1996. Green Manuring and biological Nitrogen Fixation : North Indian perspective. In Proceedings of the Symposium on plant Nutrient Needs, Supply eds. Efficiency and Policy Issues: 2000-2025 (T.S. Kanwar and J.C.Katyal), National Academy of Agricultural Sciences, New Delhi, pp 29-43.

Chapter 8

Common for the Mountains, Rare for the World:
Neglected Crops Offer Food and Nutrition Security for the Future

☆ *Silvi Siddhu*

Introduction

What is the most precarious about the global food security is this that only three food crops–wheat, rice and maize–contribute nearly three-fourths of the total energy requirements of human beings. If these crops were plagued with pest problem of epidemic proportion at a point of time threatening their production even for a single season, our very food security would be at stake. The Great Potato Famine of 1845-46 resulted from total destruction of potato crops in Peru that led to the toll of more than two million people and out-migration of another two million is one of the most talked about lessons in human history reminding us how dangerous it could be to rely on a single or just a few crop cultivars.

The opposite lesson can be drawn from the marginal lands like those in the Himalayan mountains. Mountain agriculture–from geo-political, bio-physical, and socioeconomic point of view–is marginal (Singh and Tulachan 2002). Farmers deriving their livelihoods from these marginal lands are the skilled managers of the agrobiodiversity. Himalayan mountains provide ideal niches for cultivating a number of plants that either do not thrive in the agro-ecosystems in plain areas or have been squeezed out of the farming practices in the wake of increasing production merely from mainstream food crops (mainly the wheat, the rice, and the corn). Cultivation of this diversity in the mountains is not only ecologically sound, but also avoids the chances of complete crop failure or the Great Potao Famine sort of tragedy.

Mountains' marginal farmers, despite intensive development intervention, did not cease to cultivate native crops, which, for the mainstream areas, are the so-called uncommon crops (traditionally, however, they are very much common for mountains' farming communities). This paper would throw some light on such crops. These crops, also called minor crops, marginal crops, or poor men's crops, are not only extremely sturdy, hardy and almost impervious to vulnerability of any considerable degree and, therefore, very much supportive of food and nutritional security of the marginal farming communities.

The Neglected Crops

The crops that the mountain farmers have been growing for millennia are the neglected crops in macro context. They would hardly find any significant place in policies, plans and programmes of the public sector. These marginalized and neglected crops, however, do thrive well in the agro-ecosystems in mountain areas, and that too without much care and external outputs. Their yields are considerably high. Moreover, they can be cultivated under rainfed conditions and some of the cultivars can thrive well even under very harsh conditions, including in degraded and acidic soils. They do not require much bullock and human labour (except finger millet). No chemical fertilizer or pesticide is ever used traditionally to protect these crops. These crops have been summarized in Table 8.1.

Finger millet is one of the most important crops sown during Kharif (summer) season. This crop is usually raised as a base crop mixed with so many other cultivars, like amaranth, buckwheat, and a variety of pulses and beans. Barnyard millet is at number two as far as the area occupied by this crop in the mountains is concerned. Amaranth is raised as mixed with finger millet or as a base crop along with buckwheat and kidney beans.

Table 8.1: Marginal Food Grain Crops in the Himalayan Mountains

Food Grain Crops *Vernacular Name* *English Name* *Botanical Name*	*Occurrence* *Average Yield*	*Uses of the Grains* *Common Potential*
Mandua Finger millet *Eleucine coracana*	Cultivated in wide range of habitats (from valleys to high hills) mostly in the uplands of mid-altitudes Average yield: 10-20 q/ha	Chappati (Roti); Mandua malt as complementary food for children
Jhangora Barnyard millet *Echinocloa frumentacea*	Cultivated mostly on upland mid-altitudes Yield: 10-20 q/ha	Eaten like rice; Fermentation for beer like drink
Cheena Proso millet *Pennisetum miliaceum*	Cultivated in the valleys and at mid- and high altitudes under irrigated as well as rainfed conditions Yield: 8-15 q/ha	Eaten like rice, ground for chappatis or porridge
Kangni Foxtail millet *Setaria italica*	Cultivated in wide range of environments Yield: 5-8 q/ha	Used in rituals and eaten like rice
Marsa, Cholai *Amarathus* spp.	Cultivated mostly at mid- and high altitudes Yield: 10-15 q/ha	Used in rituals, roasted and eaten in the form of laddus, ground and mixed with wheat or mandua flour for chappati making; Can be used industrially for a number of products
Ogal Buckwheat *Fagopyrum esculentum*	Cultivated mostly at mid- and high altitudes Yield: 8-10 q/ha	Used in rituals, roasted and eaten in the form of laddus; Can be used industrially for a number of products
Bathua Chenopod *Chenopodium* spp.	Cultivated mostly at high altitude Yield: 5-6 q/ha	Used in rituals, roasted and eaten in the form of laddus; Can be used industrially for a number of products

While finger millet and barnyard millet are raised on sizeable area in the mountains, amaranth and buckwheat occupy certain pockets, especially on high altitudes. Proso millet and foxtail millet are squeezed to considerably smaller areas, while chenopod for seeds is raised only in parts of Garhwal and Himachal Pradesh and rarely in other areas.

Apart from common use (made chappaties, eaten like rice, converted into flour and mixed with other food, used for making various recipes and in rituals, etc.), these food grains carry potential values also. They can be used industrially for making malt, extracting essential amino acids, minerals, steroids, medicines, etc.

Nutritional Value

As revealed from Table 8.2, the uncommon food grains have exceptionally high nutritive value. While millets are somewhat poor in the protein content, the pseudo-cereals amaranth, buckwheat and chenopod have appreciably high protein content. Moreover, protein value of the pseudo-cereals as could be exemplified from the contents of essential amino acids in it is also much superior compared to those of the common food grains (rice, wheat, maize) and even the millets. Total ash (minerals) present in all the neglected crops is very high. Finger millet among millets and amaranth among pseudo-cereals are exceptionally rich in their calcium content. The pseudo-cereals are also extremely rich in their iron content.

Quality of protein that depends on the composition of its amino acids is of critical value in human nutrition. Protein malnutrition is particularly either owing to the inadequacy or poor quality of protein or to the combination of both. Our common food grains are poor as far as the quality of their proteins is concerned. Neglected food grains of the mountain areas, however, are richer in some of the indispensable amino acids than the common food grains. Two pseudo-cereals, amaranth and buckwheat, have been compared in this regard with wheat, rice and maize in Table 8.3. Contents of various amino acids they contain are compatible with the recommendations of the Food and Agriculture Organisation (FAO) and World Health Organisation (WHO) of the United Nations Organisation (UNO).

Food Grains of the Future

The food crops raised in the marginal lands are not just traditional ones. Though they are neglected in our policy and programmes, they are the promising ones for fighting hunger,

Table 8.2: Nutritive Value of the Food Grains from Uncommon Mountain Crops

Food Grain	*Moisture, g*	*Total Ash, g*	*Crude Fat, g*	*Crude Fibre, g*	*Crude Protein, g*	*Carbo-hydrate, g*	*Energy, kcal*	*Iron, mg*	*Calcium, mg*	*Reference*
Finger millet	5.3	5.9	1.1	3.6	7.3	76.8	346	3.8	232.0	Verma and Raghuvanshi (2001)
Barnyard millet	11.9	4.4	2.2	9.8	6.2	65.5	307	2.9	0.02	Verma and Raghuvanshi (2001)
Proso millet	11.9	3.4	NA	NA	7.6	68.9	NA	5.7	10.0	Verma and Raghuvanshi (2001)
Foxtail millet	NA	3.0	6.0	8.0	9.0	74.0	NA	10.0	30.0	Verma and Raghuvanshi (2001)
Amaranth	9.3	3.5	7.1	13.40	15.3	63.1	391	22.4	490	Joshi and Paroda (1991) (2001)
Buckwheat	11.0	NA	7.4	NA	13.5	72.9	335	13.2	114	Joshi and Paroda (1991)
Chenopod	12.0	2.94	7.80	NA	14.18	62.0	395	171.0	250	Partap, Joshi and Galwey (1998)

NA: Not available.

malnutrition and for ensuring food and nutrition security for the masses. They are not just the crops of the past and the present, but also of future. Not only the people of the mountains would require them to be their indispensable partners in life, but also those dwelling in the distant plains. These food grains would be inevitably needed to be incorporated in daily diets by one and all, as these would be vital for healthy and happy lifestyles.

Table 8.3: Indispensable Amino Acids in the Uncommon and Common Food Grains (as percentage of protein)

Food Grains	*Lysine*	*Methionine*	*Tryptophan*	*Leucine*
Buckwheat	5.9	3.7	1.4	5.8
Amaranth	5.0	4.4	1.4	4.7
Wheat	2.6	3.5	1.2	6.3
Rice	3.8	3.0	1.0	8.2
Maize	1.9	3.2	0.6	13.0
FAO/WHO recommendation	5.5	3.5	1.0	7.0

Source: Joshi and Paroda (1991).

Thinking about the neglected or marginal food grains and making efforts for the development of marginal agriculture giving prominent place to such crops, therefore, would reflect progressive and futuristic perspective for ensuring food and nutrition security for communities inhabiting marginal (and even mainstream) areas.

References

Joshi, B.D. and Paroda, R.S. 1991. *Buckwheat in India*. New Delhi: National Bureau of Plant Genetic Resources.

Partap, T.; Joshi, B.D.; and Galwey, N.W. 1998. *Chenopods: Chenopodium Spp*. Rome: IPGRI.

Singh, V. and Tulachan, P.M. 2002. Marginal Farming in the Mountains. *Asian Agri-History*, 6 (3): 269-280.

Verma, T. and Raghuvanshi, R.S. 2001. *Uncommon Plant Foods of India*. Technical Bulletin Food and Nutrition II. New Delhi: ICAR.

Chapter 9

Natural Dyes for Colorful Garments and Healthy Environment: Status in Uttarakhand

☆ *Manisha Gahlot, Nargis Fatima and Neema Papnai*

Introduction

Local people in the mountains are accustomed to wear colourful garments. In order to dye their garments, they depended on natural dyes. However, ever since the modern market began to penetrate the seemingly inaccessible mountains, the local people also began to be dependent on the cloths manufactured and dyed in factories and big industries. Synthetic dyes remain the sole source to colour the cloths in the factories. Indiscriminate use of synthetic dyes has not only affected the flora and fauna but has also caused threat to human safety and thus it is a matter of immediate concern. Thus ecofriendly natural dyeing is gaining importance not only for the safety of health and environment but also for the lustrous and beautiful colors.

A dye is substrate used to colour materials in such a manner that the imparted colours seems to be a part of natural material. These compounds are normally used in solutions that are capable of being fixed on to the fabrics. They are broadly classified in to natural and synthetic dyes on the basis of origin. The natural dye stuffs are obtained from roots, stems, bark, berries, and flowers of various plants and from certain insects and shellfish.

The following properties are often considered to be the advantages of natural dyes:

- ☆ Obtainable from renewable resources
- ☆ No health hazards, can sometimes be used for health cure
- ☆ No or mild chemical reactions are involved in their preparation
- ☆ No disposal problems
- ☆ Unsophisticated and harmonized with nature

The limitations of the natural dyes that were responsible for their decline are:

- ☆ Not available in easy- to- use form
- ☆ Low colour yield
- ☆ Complex and laborious dyeing process
- ☆ Difficulty in reproducing the same shade

Classification of Natural Dyes

Natural dyes can be classified in to four categories:

1. Classification according to source
2. Classification according to colour
3. Classification according to chemical structure
4. Classification according to application

1. Classification According to Source

According to source, natural dyes can be sorted into three categories. Natural dyes obtained from plants (indigo, madder, etc.), those obtained from animals (cochineal, lac, etc.) and those obtained from minerals (ocher, umbers, etc.)

Mineral Dyes

Mineral dyes include the natural earth pigments which owe their tinctorial properties to the oxides and hydrated oxides of iron and manganese. They are insoluble in water and other solvent and must therefore be attached to the fiber surface with binders.

Plant Dyes

In these dyes the colouring matter is derived from root, leaf, trunk, bark and fruits of plants. It is believed that there are at least 300 plants from which colouring matter is extracted. Among those which have been important in the past and present, include indigo from leaves of *Indigofera tinctoria*, madder from the roots of *Rubia tinctoria*, cutch from the wood of *Acacia catechu*, henna from the leaves of *Lawsonia inermis* and safflower from *Carthamus tictorious.*

Animal Dyes

Some of the most beautiful and most expensive dyes are obtained from animals. Natural dyes can be extracted from certain insects and a few animal tissues. They include tyrian purple from glandular secretion of Perpura, scarlet from the dried bodies of *Coccus cacti* and lac from the gum lac made by the insect *Coccus laceae.*

2. Classification According to Colour

Red Dyes

The colour index lists 32 red natural dyes. The prominent among them are madder (*Rubia tinctoria*), manjisth *(Rubia cardifolia)*, brazilwood/sappanwood (*Caesalpina sappan*), cochineal (*Coccus cacti*), lac dye (*Coccus laceae)*. All these dyes are based on Anthraquinone molecule accept sappanwood based dyes. These dyes are prone to oxidation.

Yellow Dyes

Yellow is the most common colour in the natural dyes. However most of the yellow colourants are fugitive. Some of the important natural dyes are obtained from berberry (*Berberis aristata)*, tesu flowers (*Butea monosperma*), turmeric (*Curcuma lamba*), dolu (*Rheum emodi*). These dyes are flavone based dye.

Blue Dyes

The only viable choice among the blue natural dyes is indigo. Natural indigo is obtained by fermenting the leaves of various species of *Indigofera.* Woad (*Isatis tinctoria*) is another important source of indigo. The main ingredients of natural indigo are indigotin and indirubin.

Brown Dyes

Tannins are the most important ingredients in dyeing with natural dyes producing brown, grey and black colour.

Black Dyes

Source of natural black colour are alder (*Alnus glutinosa*), Custard apple (*Amona reticulata*), babool (*Acacia arebica*) and harda (*Terminalia chebula*).

3. Classification According to Chemical Structure

Natural dyes are class of colourant obtained from vegetative matter and animal residues. According to their chemical structure they can be broken down in the following categories.

Table 9.1: Categories of Natural Dyes

Colours	*Chemical Classifications*
Yellow and brown	Flavone dyes
Yellow	Iso-quinoline dyes
Orange-yellow	Chromine dyes
Brown and purple-grey	Naphthoquinone dyes
Red	Anthraquinone dyes
Purple and black	Benzophyrone dyes
Blue	Indigoid dyes
Neutrals	Vegetable tannins: gallotannins, elagitannins and catechol tannins

4. Classification According to Application

Direct or Substantive Dyes

These dyes are known as direct dyes because they colour fiber directly with one immersion without the use of a fixing agent. They are fixed on the fiber by hydrogen bonding. Direct dyes have poor light and washing fastness. However, we can improve these fastness properties after some chemical treatment.

Acid Dyes

Acid dyes are used for polyamide fibers like wool and silk. Acid dyes are applied to the substrate from acidic medium. Majority of natural dyes belonging to this class has carboxylic acid group as against the sulphonic acid group contained in artificial colourant. Colour properties of acid dyes vary widely. Some of these fade very quickly in light and from perspiration.

Basic Dyes

These are cationic dyes, which on ionization gives coloured cations. They are used on polyamide fibers like wool and silk. The dye barberine belongs to this class. Basic dyes produce brilliant colours, but they are not usually very fast to light and laundering.

Vat Dyes

Vat dyes are polycyclic compounds with chromophore in the form of a quinoid group and are insoluble in water. They are substantive towards the fibre only when they are in the form of leuco compounds which are obtained as a result of special treatment. Indigo is classical example of natural vat dye.

Dyeing with Natural Dyes

Dyeing is defined as an operation or a series of operation by means of which uniform colour of a permanent character is produced on a substrate. Since the natural dyes can not be mixed together to get compound shade. One can do sequential dyeing for getting the compound shade. However there is a limitation of shades that could be produced from natural dyes. Dyeing with natural dyes consists of three major steps namely:

A. Extraction of dye
B. Application of dye to yarn or fabric
C. Mordanting

A. Extraction of Dye

Natural dyes are extracted from different parts such as the roots, stems, leave, flowers fruits bark, seeds of the plants. The plant material are dried under shade then pulverized in the powder form. Natural dyes give colour in different medium *viz.* acidic, alkaline and neutral medium. The powdered dye can be extracted with water at neutral or acidic or alkaline pH to get the complete extraction of dye from the plant material. It is necessary to soak the dye material overnight in case of bark, roots and wood before extraction. The powdered material sources are soaked in pure water for 6-24 hours depending on its original character followed by boiling for one hour. After this process the contents are filtered so that the clear solution can be collected separately from the sludge. The boiling and filtering process can be repeated until the sludge gives the minimum residual colour. The extracted dye solution can be applied in its aqueous form or it can be converted into its powder form by evaporation process for future use.

B. Application of Dye to Yarn or Fabric

Prior to dyeing of the fiber with various natural dyes, fibers are scoured by using detergent to make them free from grease and other adhering impurities. Sometimes bleaching of the fiber is necessary to obtain clear and bright colours. Hydrogen peroxide is used as bleaching agent. The fibers are completely washed to remove all the detergent before they are dyed with various dyes. Degumming of silk is required to remove natural gum from the fiber.

For dyeing, water bath having temperature regulator are used. The fiber is taken up in a beaker containing the dye solution. The beaker is now kept in the water bath at 30-40°C temperature then the temperature of the dye bath allowed to reach up to 70-80°C with constant stirring to avoid patching. Dyeing continues for about 30 to 45 minutes. After dyeing thorough rinsing in soft water is necessary. For dyeing purpose the water should be soft and free from iron salt which saddens the colour of many dyes.

For getting the compound shades wherein more than one shade is required one can do sequential dyeing as there is a limitation of shades which can be produced by natural dyes.

C. Mordanting

The name mordant has been derived from Latin word *mordere* meaning 'to bite'. The majority of the natural dyes need a chemical in the form of a metal salt to create affinity between the fiber and the pigment and ensuring reasonable fastness of the colour to sunlight, washing and perspiration. These chemicals are known as 'mordants'. A mordant as a chemical can itself be fixed on the fiber and also make combination with the dye stuff. Mordants have direct effect on the dye and govern the colour obtained so that the same dye can produce different colours.

Types of Mordant

A. Metallic Mordants

Originally only naturally occurring metal salts were used in dyeing. Now-a-days metal salts of aluminum, chromium, iron, copper and tin are used. Some of the common mordants are: alum, ferrous sulphate, copper sulphate and stannous chloride.

Alum

This salt is also called aluminium potassium sulphate. It is most common mordant used for dyeing of silk. This mordant adds great brilliance to the colour on wool and silk without causing any damage to the fiber.

Copper Sulphate

Copper sulphate is also known as blue stone. It is generally used as a saddening agent but may sometimes be used to turn yellowish into green as the mordant tends to leave a blue haze on the yarn. By this mordant certain colours are obtained which would otherwise be impossible to create.

Ferrous Sulphate

It is one of the most important and oldest mordant known. Ferrous sulphate is easily soluble in water. This mordant is difficult to use and great care is required to ensure even dyeing. It mostly darkens the colours.

Stannous Chloride

It has a fairly high solubility in water. Like copper, it is not often used as a mordant. It is very often used as a modifying agent with other mordants. It greatly brightens the colour and helps to make them fast.

B. Natural Mordants

Mostly the natural mordant belongs to the group of tannins, which were primarily used in the preservation of leather. The tannin is the active element which is capable of combining with protein of leather to convert it in to imputrescible material. They are also used in glues, stains and mordants. Vegitable tannins are bitter and astringent substances in plants. It adds brilliance to the colour and help to keep them fast and mainly used in dyeing cottons. Common natural mordants are bahera (*Terminalia bellirica*), pomegranate rind (*Punica granatum*), tea (*Camelia sinensis*), harda (*Terminalia chhebula*), etc.

C. Oil Mordants

The main function of the oil mordants is to form a complex with alum, used as the main mordant. Treatment of oil with concentrated sulphuric acid produces sulphonated oils, which possess better metal binding capacity than the natural oils.

Mordanting Methods

There are three methods of mordanting which can be used for dyeing with natural dyes namely premordanting, simultaneous mordantig and post mordanting method.

A. Premordanting Methods

In this method the yarn/fabric is mordanted first and then dyed. An aqueous mordant solution is prepared by dissolving required amount of mordant in water. Amount of mordant should be taken up on the weight of the yarn/fabric. Boiling of yarn/fabric is done in this solution for 30 minutes.

B. Simultaneous Mordanting

In this method, the mordant and the dye are applied simultaneously in the same bath. After placing the yarn/fabric in the extracted dye bath dyeing should be carried out for 20-30 minutes. After dyeing required amount of mordant is added to the dye bath by lifting yarn or fabric and with continuous stirring. Dyeing of the yarn/fabric is done in this solution for 30 minutes. After mordanting, thorough rinsing of yarn/fabric is necessary.

C. Postmordanting

In this method, dyeing of yarn/fabric is carried out first. After dyeing of the yarn/fabric in dye solution, mordant is applied without washing the samples.

General Recipe for Dyeing with Natural Dyes

- M:L Ratio–1:20
- Acid/alkali concentration–1ml/g/100ml water
- Time for extraction of dye–30-60 minutes
- Dyeing time–30-45 minutes

- ✰ Mordanting time–30 minutes
- ✰ Amount of mordant :
  - ✰ Alum–5-10 g
  - ✰ Copper sulphate–1-2 g
  - ✰ Ferrous sulphate–2-4 g
  - ✰ Stannous chloride–2-4g

Note: Amount of mordant should be taken on the weight of the fabric

Different Steps in Natural Dyeing

Collection of dye material

↓

Selection of dye extraction medium

↓

Selection of dye material concentration

↓

Selection of dye extraction time

↓

Selection of dyeing time

Status of Natural Dyes in Uttarakhand

Recent trend are pointing towards higher preferences for natural dyes. It is supported by increased public awareness about environmental issues and actions like the German ban on use of some synthetic dyes. This has prompted and inspired researchers as well as governmental and non-governmental organizations to look at the potential of natural dyes afresh.

Uttarakhand is in advantageous position since the state holds a rich reservoir of natural products. The natural dye sources are abundantly available in villages and forests of Uttarakhand. Different parts (Leaves, bark, seed, flowers, roots and wood) of number of plants have been reported to yield the dye but a large number of them hitherto remain unexplored. Therefore, there is a need to explore and utilize these dye sources as a means of income generation. The rural people can be trained in identifying the dye yielding plants, collecting and storing the dye material. They can generate additional income through selling the dye material in identified markets. Many of the vegetable dye sources are herbs, shrubs and other plant matter that could be cultivated in dry lands. This leads towards effective utilization of wasteland and generate rural employment.

Uses of Natural Dyes

Natural dyes have wide scope in fashion industry even though there is restricted colour palette. Their colour variation is an exciting experience to designers as there is flexibility in use of natural dyes. We have unlimited resources at our hand in terms of fibre and yarn dyeing with natural dyes for weaving and carpet industry, in hand block and screen printing for surface decoration and dyeing of fabric with natural dyes. Natural dyes give scope for providing part time employment for rural masses. These dyes can easily be adopted by the weavers, artisans, women groups and handicraft makers to produce goods of consumer satisfaction. Thus cultivation of dye yielding plants and trees will be an agro-based activity in rural areas leading to rural employment. Another area to improve the scope of application of natural dyes is by production of colour concentrates in either paste or powder form.

Natural dyes can be used for preparing herbal dry colours (*gulals*). Herbal hair colours can also be prepared by the use of natural dyes. All of the dyes have therapeutic uses as botanicals (herbs), or as homeopathics (diluted), or as nutritional supplements (minerals). Natural colours are also being used in cosmetics, food and pharmaceuticals.

Conclusions

Natural dyes are produced by nature, but that does not mean that the entire natural dyes are ecofriendly. If biodiversity is disturbed it can not be considered ecofriendly. So if we cut one tree to use the dye stuff, we should grow at least three new trees to replace the one we have used. Waste material such as onion beels should be utilized to overcome these problems. We can also try to start growing some of our own dye stuff as it is a great fun and rewarding.

References

Sekar, N. 1999. Natural dyes: prospects and future. In: *Convention on Natural Dyes.* Delhi, December 9-11, 1999, Delhi IIT. p37

Gulrajani, M. L. 2001. Technology for production and application of natural dyes on textiles. In: *Convention on Natural Dyes*. Delhi, December 17-18, 2001. Delhi IIT. p3.

Lyengar, M.M.S. 2001. Research Briefs. *Indian silk*, 47. 4:4.

Chapter 10

Traditional Knowledge System of Himalayan Dyeing Techniques: Cultivation of Dye-yielding Plants as a Contribution to Eco-development

☆ *Saurabh*

Natural dyes were extensively used for the dying of all natural fibers until the middle of the 19th century. The discovery of synthetic dyes alizarian and indigo leads to death of indigenous natural dyeing industry. The ease of application and availability of synthetic dyes drove the art and craft of the natural dyes into oblivion. Excessive use of the synthetic dyes and the non-treatment of effluents contained in the waste water of the dying process have led to horrendous results.

Comparison of Natural Dyes and Chemicals

The history of natural dyes is more than 4000 years old. Certain advantages synthetic dyes have over natural dyes like good repeatability of shades and brilliance in color performance as they provides multiple color with bright tint.

Present Trend

Nowadays with the environmentally benign products, becoming top priority, fashion industry has turned to produce newer products which are according to fashion trends and eco-friendly as well.

Natural dyes being based on saw material like roots, leaves, flowers, fruits bark insects, mineral, etc., available in the nature do not create any pollution problem rather in some cases like harda indigo etc, the "Waste" in the processes becomes an ideal fertilizer for use in agriculture fields.

The global consumption of textiles is estimated at around 30 million tonnes, and this is expected to grow at 3 per cent/annum. The dyeing of this much quantity of textiles needs around 7,00,000 tonnes of dye stuff. Thus it is immense importance of find the additional sources of natural dyes which are available in plenty. There is growing trend all over the world for the natural colorants and the present control context of eco preservation natural colorants, have got tremendous commercial potential as there is also a growing awareness of their non-hazardous nature.

In recent years attempts have been made to revive the art of vegetable dyes due to the carcinogenic nature of synthetic dye and intermediaries.

Our Heritage

India was a major producer of natural dyes in the past and the natives have always extracted days from the local plant materials in every part of country. In Rajasthan and Andhra Pradesh artisans are still using natural dyes for tie and dye, block printing, Kalam Kari and wall decoration.

In Uttarakhand

Central Himalayan region have various ethnic communities viz-*Bhotiya* tribe make a range of woolen garments and materials, besides processing, coloring of wool and clothe. Many plants bearing dye pigments are available in Himalayan range.

Originally the indigenous knowledge of making natural dyes from nature plants species had developed in the high altitude region of Kumaun over a long period of time. Dyeing of wool with natural dyes was very common in Pithoragarh in Kumaun and Chamoli and Uttarkashi districts of Garhwal. Even these people were so perfect that they could get most of the bright shadese of colors required in dyeing of woolen.

Origins of Natural Colorants

Animal Origin

Lac, cochineal, Kermes, etc.

Mineral Origin

Prussian blue, Red ochre, Ultramarine blue.

Vegetable Origin

Pigment derived from roots, bark, leaves, stem, fruit etc. of plants.

Table 10.1: Some Vegetable Origin Colors Produced and Used in Himalayan Region

Sl.No	*Color Produced*	*Local Name of Plant*	*Botanical Name*	*Part Used*
1	Orange	Harsingar	*Nyctanthes arbortristis*	Flower
		Rudhi	*Mallotus philippinensis*	Fruit
2	Violet and	Jamun	*Sygygium cumini*	Fruit
	Blue shades	Makoi, Ninoni	*Solanum nigrum*	Ripe. fruit
3	Black	Darim	*Punica grantum*	Fruit shell
		Urad	*Vigna mungo*	Seed
		Ratgali	*Abrus precatorius*	Seed
4	White	Sanar	*Ougenia oojeinensis*	Stem
		Genhu	*Triticurm aestivum*	Grains
		Chawal	*Oryza sativa*	Grains
		Gethi	*Boehmeria rugulosa*	Stem
5	Green	Jhatalir	*Principia utilis*	Leaf
		Makoi, Ninoni	*Solanum nigrum*	Unripe fruit
		Datura	*Brugmansia suaveolen*	Leaf
6	Yellow	Kilmora, Kingod	*Berberis* spp.	Root
		Dolu	*Rheum australe*	Root
		Archa	*Rheum moorcroftianu*	Root
		Khukhuyinya	*Rumex nepalensis*	Root
		Dholi	*Wood fordia floribunda*	Flower
		Palaas	*Butea monosperma*	Flower
		Hajari	*Tagetus* spp.	Flower
		Lodh	*Symplocos crataegoide*	Bark
		Basing	*Adhatoda vasica*	Leaf
		Bel	*Aegle marmelos*	Root
7	Red	–	*Geranium nepalense*	Root
8	Camel	Akhrot	*Juglans regia*	Fruit
		Kapasi	*Corylus jacguamontii*	Fruit shell
				Fruit shell

Future Perspectives

There are many other plants, which have to be identified for dye extraction. In Himalayan region mostly every plant have some colorant. Hence attempts have been made to identify combinations to produce more shades of colors that are already identified like range of browns, yellow, black, and to standardize/optimize their dye application precedence.

In spite of several advantages natural dyes colour is still very limited due to non-available of standards, shade cards precise and specific ways of application and standard norms. The appropriate technical knowledge of color extraction, purification and standardization of dyeing processes is of immense value and requires detailed study. Thereafter only use of natural dyes becomes comparable to synthetic dyes.

Natural Dyes in Rural Development

Natural and vegetable dye material cultivation, collection processing, dye extraction and dyeing of textiles are considered to be agro based rural industry helps promote rural entrepreneurship, particularly in Khadi, Wool and Handloom centers. Thus developing rural economy.

Cultivation of vegetable dyes is one of the agro based rural industry. Many of vegetable dyes source of plant matters could be cultivated in dry land and this would lead to effective utilization of wasteland and generate rural employment. Cultivation of dye matter plants gives dual benefits, as it gives dye material as well as fuel wood for rural people.

References

Book of papers: Conventions on Natural Dyes. IIT, DELHI

C.P. Kala. 2002. *Current Science*, 83(7): 815.

Chapter 11

Unconventional Fibres: Processing and Economic Importance

☆ *Manisha Gahlot, Neema Papnai and Nargis Fatima*

Introduction

Indian textile industry has retained its importance in the economy of India from the foreign exchange earning point of view. It contributes over 20 per cent of total national industrial production, and over a third of national exports. Since last decade, the situation has not remained the same because of the competition from China. It is realized that for survival of the Indian textile industry and also for improving the export of these goods, certain diversified value added products such as handicrafts are needed to be developed. The textile industry has close linkages with the agricultural sector and is dependent for its raw material on the output of millions of small farmers and growers. India has long traditions and extensive experience in the fibres group and in textile related applications.

Plant fibres, in fact, are the raw material for many vital industries sustaining the economy of our times. In the processing of these fibres an estimated 20 million people are involved, in production, post-harvest management and primary processing, in the decentralized cottage and small industry segments and in the relatively organised mill and power loom sectors. Natural fibres of vegetable origin are basically cellulosic in nature and have varying amount of associated lignin and pectic substances in them. Interestingly, about 90 per cent of the natural fibres being produced are of vegetable origin. Cotton constitutes nearly 80 per cent; the other 20 per cent of long vegetable fibres are made up by jute, flax, hemp, sisal, ramie, coir, abaca, and pineapple. Value added handicraft products are prepared from these fibres to sustain the economy of the industry. The economic importance of handicrafts is accentuated by the fact of high employment potential, low capital investment, high ratios of value addition and that it is almost wholly artisan based.

The handicrafts sector also employs a large number of people, many of whom work with traditional materials, processes and products, and in remote and backward areas. Government of India estimates of direct employment in this sector at about 6.5 million people in rural and urban areas, approximately a tenth of the total labor force in the manufacturing sector; women are about 60 per cent of this work force. An indication of the importance of this sector in India's macro-economic strategy is the fact that it accounts for about 6 per cent of exports, valued in 1995-96 at about Rs. 5,000 crores. In the 1990s, handicrafts exports have grown at an unprecedented growth rate of about 30 per cent annually, reaching a figure of Rs. 5660 crores in 1996-97. India's Ninth Five Year Plan identifies this as a thrust sector, to create higher employment opportunities and a better standard of living for craftsperson individually and collectively through efficient production practices, professionalized management, enlarged marketing networks, for the domestic market and overseas, and preservation of the rich cultural heritage of the country.

Natural Fibres

An abundance of fibrous material occurs in nature and over the centuries man has experimented whether these natural fibres could be used as textile fibres for various end uses or not. These fibres can be grouped under two categories, *viz.*, fibres which are cultivated because of their utility and commercial importance such as cotton and those which are naturally grown. Natural fibres are of many types, which depend upon the plant from which the fibre is obtained. These vegetable fibres may also be classified as seed fibres, leaf fibres, fruits or nut fibres, stem, bark or bast fibres:

1. *From seed fibres*: Cotton, kapok, milkweed floss.
2. *From stem or bast fibres*: Flax, hemp, ramie, jute, hibiscus, bichua.
3. *From leaf fibres*: Pineapple, sisal, manila, murva, rambans. (agave).

Table 11.1: Classification of Vegetable Fibres

Class of Fibre	*Fibres*	*Botanical name*
Bast fibres	Jute	*Corchorus capsularis, olitorius*
	Flax	*Linum usitatissimum*
	Ramie	*Boehmeria nivea*
	Hemp (Bhang)	*Cannabis sativa*
	Sunn	*Cannabis juncea*
	Kenaf	*Hibiscus cannabinus*
	Urena	*Urena lobta*
	Nettle	*Urtica diocea*
	Jungli bhindi	*Hibiscus feculneus*
	Musk mallow	*Hibiscus abelmanschus*
	Bichua	*Girardinia heterophylla*
Leaf fibres	Banana (abaca)	*Musa sapientum*
	Agave (sisal)	*Agave sislana*
	Furcrarea(Mauritius)	*Furcraea gigantea*
	Murva	*Sansevieria trifaciata*
	Pineapple	*Ananas comosus*
	Rambans	*Agave americana*
	Henequen	*Agave fourcroydes*
	Phormium	*Phormium tenax*
Seed and	Coir	*Cocos nucifera*
miscellaneous fibres	Tree cotton	*Malvaceae species*
	Kapok	*Ceiba petandra*
	Milkweeds	*Asclepias species*
	Kumbi (galgal)	*Cochlospermum gossypuim*
	Cotton	*Gossipium hirsutum*
	Semul	*Salmalia malabarica*

Processing of Fibres

The processing of unconventional fibres is of great importance since the quality of extracted fibres is strongly influenced by the method employed for getting the fibre. The various steps involved are as follows:

Harvesting of Plant

Part of the plant or fibres are pulled by machine or by hand.

Fibre Extraction

All methods of fibre extraction have the aim of removing the fibres which are in the stem sheaths from the surrounding flesh. Besides this, fibre products can also be produced directly from different parts of plants.

(*i*) Bacteriological Processes

The term bacteriological processes, which are also called retting processes, describes the extraction of the fibres from the plant by bacterial retting. The main point in retting is that favourable living conditions for bacteria have to be provided, the main precondition being ample moisture. The following retting processes are distinguished:

Dew Retting

For dew retting the fibre sources are spread out on damp ground so that fungi and probably also bacteria can expose the fibres within a period of one to two months.

Water Retting

In water retting, there are a number of different methods, depending on the temperature (*e.g.* cold and warm water retting), the location of the water (*e.g.* ditches, ponds, rivers) and the bacteria (*e.g.* anaerobes and aerobes).

Allowing the retting processes to continue too long will deteriorate the quality of the fibres. On completion of the retting process, the fibres are taken out of the water and other types of loose cellulose matter are removed by wiping and washing manually.

(*ii*) Manual Extraction

Extraction of fibres by hand is a very old technique which basically consists of knocking or scrapping the fibres out of the stem or leaf flesh. The most common method is to pull the sheaths through two knives pressed together, in order to scrape the flesh off the fibres.

(*iii*) Mechanical Processes

The basic principle of the most common machines is that the individual sheaths are passed between rotating rollers, in order to reduce the high percentage of moisture and expose the fibres. These machines consist of rotating blade roller driven by a motor, and a shoulder plate which is arranged at a tangent with a small gap, to the blade roller. The individual sheaths are passed between the blade roller and shoulder plate into the machine. The blade roller scrapes off the leaf flesh and exposes the fibres.

Breaking and Scutching

The next stage in fibre production is breaking. The straw is passed between fluted rollers in a breaking machine, so that the woody core is broken into segments without damaging the fibres running through the stems. In scutching process the broken pieces of stems are knocked off.

Hackling

After scutching or retting the fibres are usually combed or hackled by drawing them through sets of pins, each successive set being finer than the previous one. The coarse bundles of fibres are separated into finer bundles, and the fibres are also arranged parallel to one another.

Finishing of Fibres

Most natural fibres are endowed with, or have embodied in them, certain amounts of extraneous impurities at their source which prohibit immediate use in yarn manufacturing. In the first category fall such physical impurities as leaf, twigs, burrs, sand, seed particles, motes, bark, and neps (tangled fibres). In the second category there are waxes, oils, greases, natural gums and lignin. Some are easily and quickly removed and some adhere tenaciously and require chemical treatment. Various finishing processes such as mercerization, scouring, degumming, bleaching and dyeing can be employed on fibres.

Mercerization

For mercerization the fibres were combed and made into small bundles. It was then dipped into 18.0 per cent NaOH solution at

30°C for 1 hour with a fibre to liquor ratio 1:20. The bundles were then washed with water, scoured with acetic acid and again washed with water until free from acid and then air dried.

Scouring

The samples were boiled in the scouring solution (Caustic Soda 2 per cent, Soda Ash 2 per cent, Turkey Red Oil 0.5 per cent, Soap Oil 0.5 per cent, M:L ratio 1:20) for six hours. Then the samples were squeezed, rinsed thoroughly in hot water and twice in cold water. The treated samples were shade dried.

Degumming

Chemical degumming of fibre is done by subjecting the fibre with solution of acids and alkalies at varying levels of temperature and duration of treatment. Sodium hydroxide is used as the main chemical. For effective degumming the fibre samples are treated with 1.0 per cent NaOH solution with the liquor ratio 1:6 or 1:7 and containing some wetting agent at 96-98°C for 2 hrs. Addition of sodium sulphate to the extent of 0.5 per cent in the degumming solution improves the tenacity of the fibre. If the degummed fibre is again subjected to 1.0 per cent alkali treatment at 120-125°C for 2 hours, the feel of the fibre and fibre separation become better.

Figures 11.1: Fibre after Retting

Bleaching

Bleaching of fibres was carried out in an open bath (M:L ratio 1:50) for 50 minutes at 90°C with hydrogen peroxide (4g/lit), trisodium phosphate (5g/lit), sodium hydroxide (1g/lit) and sodium silicate (10g/lit) as bleaching bath assistants. The bleached samples were washed with cold water and then neutral wash was carried out with 2 g/lit acetic acid. It was again washed and dried. Only a mild bleaching treatment is required. However special care is necessary to prevent formation of oxycellulose in bleaching.

Dyeing

Direct, azoic, sulphur and vat colours are mostly used. Direct colours with good light fastness are used for furnishing fabrics. These fibres are harder than cotton, and dyestuff do not penetrate so readily into it. Special techniques are used during dyeing to ensure maximum penetration of dye, such as the pigment padding process (vat dyeing).

Product Development

Techniques Employed

Spinning

Spinning may be defined as the ease with which textile fibres may be twisted into a continuous, uniform yarn, having commercially acceptable properties and cost.

Fabric Formation

Weaving

In weaving, some sort of frame loom is required on which yarn can be stretched lengthwise to form what is known as the warp. After warping another set of yarns (or other material) are passed over and under the warp widthwise to form the fabric. The frame loom can be constructed out of an old picture frame, a canvas stretcher or four strips of wood. Just nails are required on which to stretch the warp yarns. The loom can be made small enough to hold in the lap, or of a larger size that will stand against the wall.

Off Loom Weaving

Knitting, Braiding, Macramé, Crocheting etc. come under off loom weaving technique.

Knitting and Crocheting

Crocheting and knitting is a method of construction a firm or solid structure without depending on a backing foundation. Only a few basic stitches form the structure. The basic difference in crocheting and knitting is that in crocheting only one needle makes the structure while in knitting two or more needles are used.

Braiding and Macramé

Braiding and macramé are true hand crafts. The fingers are the main tools and use of all other equipment is incidental. The structure of braid and khats of macramé determines the design which can be primarily a simple one made with colour and texture placement. These are fast rhythmic method of producing a sturdy, long lasting and hard wearing structure.

Sheet Making

In this technique fibres are aligned parallel on a piece of fabric and stitching is done at regular intervals of 1-2 inches.

Non Woven

Web Formation

Another utilization of fibres (both staple and continuous filament) is to convert the fibres into a sheet material through various processes used in web formation.

Bonding

The fibres have to be held together either by entangling them or by incorporating a bonding agent such as resin, solvent, or a polymer melt. Bonding techniques are of following types

Chemical boding- resins are used to bond fibres.

Thermal bonding- fabrics are made by heat setting.

Stitch bonding- the needle passing through the subjected layers of material to their maximum penetration position.

Products and Uses

Handicrafts Products

Value added handicraft products can be made with the use of these fibres. Shopping bags, pot holders, magazine covers, hat, purses, wall hangings, pen kits, mats and basket are some of the examples of such products.

Utility Products

These fibres also find use in cordages for making small domestic ropes, parceling twines, and horticultural twines.

Woven Fabrics

These fibres can be used in blends with other fibres or as weft in weaving.

Non Woven

Durable non woven fabrics include material such as interlining and interfacing, automobiles interiors, hood insulator, luggage, furniture, furnishings (mattress covers, under spring covers, draperies etc.), carpet backing (primary and secondary), wall coverings, fibre fill (pillows, mattresses, sleeping bags and life preserves etc.), stuffer material in car seats, roofing tiles, architectural fabrics, suede like composite structures, and geo textiles.

Table 11.2

Fibre	*Products*	*Technique Employed*
Rambans	Towel hangers, hand purse	Macramé
Bhang	Coasters, basket, table mats	Off loom weaving
Murva	Mats, waist belt, rugs	Çoiling, macrame
Bhindi	Cords, doormats, bags, belt	Braiding, macramé, twisting
Hemp	Fabric, canvas	Weaving
Kenaf	Twins, cords	Braiding
Jute	Bagging, wrapping, padding cloth, carpet, rugs	Weaving, tufting
Flax	Lace, bags, clothing	Weaving, lace making
Sunn	Fishnets, paper	Knotting, pulp making
Nettle	Cordage, sail cloth	Weaving, twisting
Cotton	Clothing	Weaving, knitting
Banana	Shopping bags, pencil bags	Braiding, sheet making

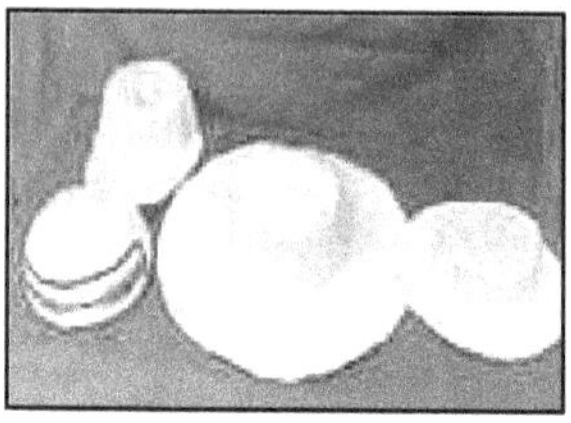

Hats

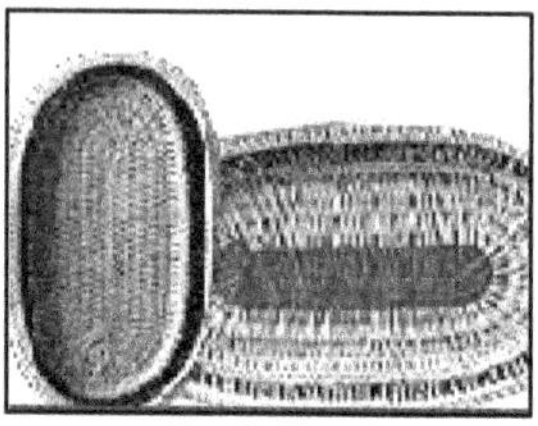

Fruit Basket

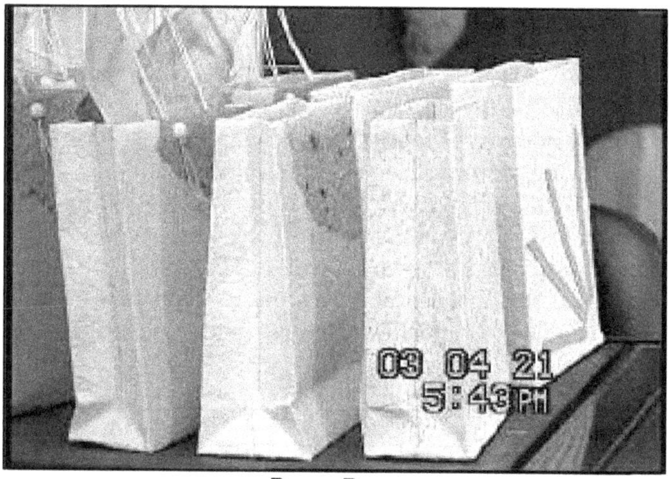

Paper Bags

Sheets

Figure 11.2: Products made from Natural Fibers by Using Various Techniques

Industrial Application

While most fibres are used in the manufacture of fabrics and clothing, yet they have numerous other uses. For instances, the electrical industry employs fibres with high insulating qualities, either inherent in the fibres or subsequently imparted to them. Paper industry also utilizes these fibres as raw materials for making paper.

Suggestions

The main problem facing this industry is lack of availability of good quality raw material. Special emphasis should be given on improving availability of good quality raw material, its processing and grading to mobilize institutional finance for development of this industry. Design improvement is yet another factor which needs special consideration and attention. Today's consumer demands quality and innovation in the product. Also, proper arrangements for strengthening marketing of finished products should be accorded due priority. For the marketing of handicraft products, action should be taken to develop outlets in tourist centres, and to provide increased access to markets. Realising the fact that marketing has a crucial role in development of rural non-farm sector and rural *haats* being an integral part of a rural market system, these should be promoted.

For revival of this industry Government should play the role of a facilitator in ensuring adequate backward and forward linkages for the growth of these traditional industries, *i.e.* availability of raw materials, training of artisans, identification of new markets etc. Training by Master Craftsmen should be organised in situ and these locations should be developed as "Shilp Grams". In addition, assistance should be sought under the Government of India's "Baba Saheb Ambedkar Hastshilp Vikas Yojana" which aims at developing the Handicrafts sector through an integrated cluster approach and by making the artisans self-sustainable. Assistance to these clusters should be provided through establishing suitable linkages with on going Central and State sponsored schemes/programmes.

Although these fibres are abundantly available, there has been little advancement in technology, product diversification and development. Activities under the schemes and programme should also seek to promote the use of craft output as an ecofriendly and environmentally sustainable alternative. A key element should be the setting up of common facility centres in craft concentrations, to enable access to inputs by artisans. Envisaged assistance should focus on selected craft related activities, with an emphasis on technological and design inputs to enhance productivity, income and employment, to facilitate economic and social upliftment of the crafts persons. The byproducts can also be utilized for making medicines, organic manure etc. The export of these fibres is one of the main sources of foreign exchange for the country. These fibres if utilized beneficially can be a good source of income, as well as an oppourtunity of employment to the unemployed.

References

Maueresberger, H.R. 1954. *Mathew's Textile Fibres*. New York, John Willey and Sons Inc. pp.

Natrajan, K. Thangamani and Thilagavathi, G. (eds.). 1997. *Proceedings of AICTE-ISTE Summer School on Unconventional Natural Fibres and their Textile Applications*, PSG Textile Technology, Coimbatore, 1997.

Kaur, S. and Gahlot, M. 1997. Agro based fibers. *Textile Industry and Trade Journal*, 35(3): 31-33.

Chapter 12

Preserving Indigenous Belief System of Dev Bhumi Uttarakhand: A Great Tradition of Fairs and Festivals

☆ *Shachi Shah and Mrinalini Shah*

Introduction

Himalayas are known as the abode of the gods since times immemorial. Despite the vicious onslaught of modernization, Uttarakhand the 26th state of India, still gives us deep insights into what authentic learning societies could look like. Their indigenous belief systems, local knowledge systems, festivals and social relationships comprise diverse and rich learning spaces, rooted in and evolving with local culture and environment. The people share a deep and personal relation with deities and identify themselves with them at different levels. As a result, these deities are a part of and not 'apart from' their daily life cycle. For the local people, their village societies offer opportunities for them to identify challenges and provide solutions, which are relevant to their local context and consistent with their values of interdependence, diversity, cooperation and collective participation. Their essential human values and how they manifest themselves in villages. The great tradition is enveloping the little one. The local beliefs, rituals, deities are now under the umbrella of great tradition. The tribal beliefs and practices are under transformation. The high morality and ideals taught by Ramayana have been constantly persuading the people to shun evils and proceed on the path of virtue and goodness. Out of this deep relationship with the land emerge several diverse activities and traditions that reinforce a holistic sense of identity, community relationships and place. Folk songs of Uttarakhand, which narrate everything from the community's historical events and personal tragedies surround nature. Embedded in these songs are individuals' and collectives' vast and detailed knowledge about agricultural practices and nature itself.

The religious, social and cultural urges of the people of Uttarakhand find an expression in various fairs and festivals, which are in turn closely linked to the economic activities of the region. Various folk songs and dances have been kept alive because of these fairs. In older times, when means of transport were not so good these fairs happened to be an opportunity for friends and relatives to meet regularly. In spite of all the change in our society, the people have kept the tradition of fairs. These festivals reaffirm the strong conviction that individual good can only result from collective good. Spaces and processes are centered on land, forest and agricultural practices, because these are the sources of collective well being. They are a source of pride and inner strength for individuals, and they also emphasize interdependence, intrinsic motivation and collective participation among members of community.

Identity and History

Uttarakhand finds mention in the ancient Hindu scriptures as Kedarkhand, Manakhand and Himavat. The Kushanas, Kudinas, Kanishka, Samudra, Gupta, the Pauravas, Katuris, Palas, the Chandras and Pawaras and the British have ruled in turns. Uttarakhand, formerly a part of Uttar Pradesh, was formed on November 9th, 2000 as the 27th state of the Indian Union. The present state of Uttarakhand was earlier a part of the United Province of Agra and Awadh, which came into existence in 1902. In 1935, the name of the state was shortened to the United Province. In January 1950, the United Province was renamed as Uttar Pradesh and Uttarakhand remained a part of Uttar Pradesh before it came into being on 9 November, 2000, the 27th State of India.

Uttarakhand lies in the Northern part of India amidst the magnificent Himalayas and dense forests breathtaking in its natural splendor, grand in its wild denizens, and simplicity in its people. The state is bordering Himachal Pradesh in the north-west and Uttar Pradesh in the South and has international borders with Nepal and China. The State today with 13 districts can be grouped into three distinct geographical regions, the High mountain region, the Mid-mountain region and the Terai region. Dehradun is the capital city. Uttarakhand consists of 13 districts *viz.*, Pithoragarh, Almora, Nainital, Bageshwar, Champawat, Uttarkashi, Udham Singh Nagar, Chamoli, Dehradun, Pauri, Tehri Garhwal, Rudraprayag, and Haridwar. It is one of the most beautiful and enchanting regions of northern India. Nature has endowed this land with so much beauty and spiritual bliss that it is also known as *Dev Bhoomi*, the Land of Gods. *Ganga, Yamuna* and scores of other rivers originate in Uttarakhand. Among them Ganga is the most holy and prominent as she represents the soul of India–her rich culture, history and civilization. In *Sanskrit* 'Himalaya' means Abode of Snow, truly characterizing the vast permanent snow fields above the snow-line. In the heart of these majestic mountains lies the state of Uttarakhand with Kumaon region in its east and Garhwal in the west.

The Garhwal Himalayas have nurtured civilization from the wee hours of history. It appears to have been a favorite locale for the voluminous mythology of the Puranic period. The traditional name of Garhwal was Uttarakhand and excavations have revealed that it formed part of the Mauryan Empire. It also finds mention in the 7th-century travelogue of Huen Tsang. However, it is with Adi Shankaracharya that the name of Garhwal will always be linked, for the great 8th-century spiritual reformer visited the remote, snow-laden heights of Garhwal, established a *math* (Joshimath) and resorted some of the most sacred shrines, including Badrinath and Kedarnath.

Following chiefly the agrarian-pastoral way of life, native Garhwalis make their living from the hilly land the best they can. Some, like the Bhotia traders, migrate far and wide, although the ancient trade routes with Tibet have been closed since 1950. Most of the indigenous people like Jaunsari, Bhotia, Buksha, Tharu, and Raji are heterodox Hindus and Buddhists, while Sikh migrants from West Punjab have settled in the lowlands since 1947. A few Muslim groups are also native to the area, although most of them have settled recently. The Muslim Gujjar herders also migrate to the hills.

The early medieval history of Kumaon is the history of the Katyuri dynasty. The Katyuri kings ruled from the seventh to the 11th century, holding sway at the peak of their powers over large areas of Kumaon, Garhwal, and western Nepal. The town of Baijnath near Almora was the capital of this dynasty and a center of the arts. Temple building flourished under the Katyuris and the main architectural innovation introduced by them was the replacement of bricks with hewn stone.

The inhabitants of the Kumaon hills are commonly known as the Kumaonese. They belong to a predominantly patriarchal society, which recognizes the superiority of men over women. The social structure is based on the extended family system, the eldest male member being the head of the family. Women are respected in society but they usually confine themselves to household activities. No religious ceremony is considered complete without the wife joining the husband. Women also work in the fields and forests alongside the men. In Kumaon, bigamy or polygamy has religious sanctions but polyandry is prohibited. Widow marriage and re-marriage by divorced women is permitted except in certain sections of high-caste Brahmins and Rajputs. Child marriage is still practiced in all classes of society despite legal restrictions on it. It has all but disappeared from the educated society.

On a hilltop facing east (opposite Almora), is the temple of Katarmal. This 900-year-old sun temple was built during the declining years of the Katyuri dynasty. The intricately carved doors and panels have been removed to the National Museum in Delhi as a protective measure after the 10th-century idol of the presiding deity was stolen.

After an interregnum of a couple of centuries, the Chands of Pithoragarh became the dominant dynasty. The Chand rulers built the magnificent temple complex at Jageshwar, with its cluster of a hundred and sixty-four temples, over a space of two centuries. Dedicated to Lord Shiva, the evocative carvings are complemented by the beautiful cedar forest around it.

Uttarakhandi Folk Songs

The Himalayas have inspired generations of singers, balladeers, and musicians throughout the ages. The natural beauty of the mountains–which inspires a deep spirituality–and the harshness of life–which darkens the heart with adversity and anguish–have invigorated Uttarakhandi music, heightening its poignancy and enriching its lyrical texture. This has been especially true of the *Pahari* penchant for folksy values, biting humor, raw honesty, loving devotion, and soaring spirits that shines through each song. Indeed, in face of other homogenizing trends in India, Uttarakhandi music has distinguished itself in retaining its cultural uniqueness and vibrant sound, while remaining popular among the masses.

There are songs about the various changes in a woman's body during pregnancy, about mythological, social and historical events. These songs are critical not only for keeping communities' histories alive, but also for providing spaces for women to express themselves and nurture their creativities. Chhopati is the love songs sung between the men and women in the form of questions and answers. Chounphula and Jhumeila are songs form part of seasonal dances, which are performed from Basant Panchami to Sankranti or Baisakhi. Jhumeila is sometimes mixed but is usually restricted to women. Chounphula is a spinning dance performed by all sections of the community at night in-groups by men and women. As the name suggests, Basanti songs are sung during the approaching *Basant* (Spring) season. Mangal Mangal songs are sung during marriage ceremonies. Jaggar is sung during worship and in respect to God Bajuband Love song sung by shepherds in form of conversation between man and woman. Khuded song is sung to portray the suffering of a woman due to parting from her husband. Chhura folk songs are very popular among shepherds and are in a way of an old experienced man teaching a young shepherd the tricks of his trade

Uttarakhandi Folk Dances

Bhotiya Dance is performed by Bhotias and is connected with death ceremonies. They believe that the soul of the dead person is residing in body of a goat or sheep and by dancing the soul can be liberated. Barada Nati dance is performed during religious festivals or any other social occasions. Men and women wearing colorful costumes perform it. Pandava Dance is performed by narrating the story of Mahabharata accompanied by dance and music. This is performed during Dussehra and Deepawali.

Langvir Dance is an acrobatic dance performed by men. In this dance the dancer climbs a pole and balances himself at his navel on the top of the pole. Accompanied by dhola and music he balances and rotates on his belly at the pole and performs other acrobatic stunts.

Major dance forms of the region are Langvir Nritya, Barada Nati folk dance, Pandava Nritya, Dhurang, and Dhuring

The Kumaonese are fond of music, folk dance, and songs accompanied by local musical instruments like murli, bina, and hurka. The hurka is played by the "jurkiya" and the dancer accompanying him, known as "hurkiyari," is usually his wife or daughter. They go from place to place narrating folklores, singing the praise of their gods and goddesses. During fairs and festivals and at harvest time, the Kumaonese often dance the Jharva, Chandhur Chhapalior, and many other forms of folk dances. The popular folk songs are Malushahi, Bair, and Hurkiya Bol.

Uttarakhandi Festivals

There are numerous festivals that celebrate the link between community well being and nature. A good harvest is celebrated with gratitude towards the power that has created abundance in nature. Benediction or *kripa* is vital for survival, and most of the festival rites are associated with offerings of the new harvest to the local gods and goddesses. Each festival is related to a particular agricultural activity, which in turn is related to time, seasons and systems of measurement. As such, they link time, movement of the earth and shifting of the sun's position with agriculture, food and health, and more broadly to peoples' own faith, gratitude and ultimately happiness. Such festival traditions both humble and empower community members, by celebrating forces greater than himself/ herself and asking for their protection and guidance. They also provide excellent opportunities for people to heal and replenish different relationships.

The festivals often involve much preparation and several rituals. Several of the seasonal festivals observed in the region (such as Harela, Aatthon, Khataduva, Uttarayani, Basant Panchami, Phooldei and Ganga Dussehra) seem to have scientific basis. The Harela festival (also called variously Hariyala or Hariyali, the green festival) celebrated on the first day of traditional month of Shravan (July-August) symbolizes mutual harmony of man and nature as well as occasion to formally wish for the successful harvest of the kharif crop as working the land requires collective effort, the festivals also call for everyone's participation. On *Phooldei*, in the months of March/April, young girls put the first flowers on the threshold of every house in the village, for good luck throughout the year. During July/August, the men celebrate *Pooja* by lighting a lamp in the village temple and taking turns in keeping guard to ward off evil spirits. If things go wrong in the village (such as illness or tensions) then the people of the village collectively do *Pooja*, by offering prayers to appease the angry *Dharti-Mata* (Mother Earth). The rituals or prayers involved during these festivals are for the well-being of all the households in the village and never limited to the well-being of "me and my family alone".

The significant learning from these celebrations is their collective nature and generosity of spirit. There are also many festivals, ceremonies and rituals that mark the different events of life: birth, marriage, and death. For such occasions, after the individual activities of bathing, cleaning homes and courtyards, and wearing clean or new clothes, food is shared with the rest of the community, especially the less fortunate. Thus, festivals are times to learn *barkat*, which literally means 'plenty', but also denotes 'generosity of spirit' and 'social responsibility', not simply towards the family but within the whole community.

Fairs

Numerous fairs and festivals are celebrated all over Uttarakhand every year. In the Kumaon hills, Uttarayani Mela is held at a number of places including Bageshwar, Rameshwar, Salt Mahadev, Chitrashila (Ranibagh), Pancheshwar, etc. on Uttarayani day. At Pancheshwar the dola of Chaumu comes down to the temple. Nanda Devi Mela is celebrated throughout Kumaon for prosperity, both material and spiritual. Nanda Devi Melas are held at Almora, Nainital, Nauti, Dandidhara, Munsyari and Ranikhet, of which the Almora and Roopkund fairs are the most famous. At Almora, thousands of people take part in a procession carrying the *dola* of Nanda Devi. The Nanda Devi temple of Almora built during the reign of Raja Udhyot Chand, is the venue for the three day long Nanda Devi Fair. Of great religious and cultural significance, it is held at Nanda Devi Temple in September to commemorate the memory of Goddesses Nanda and Sunanda. It is said to have started in Kumaon during the reign of Raja Kalyan Chand in the 16th Century.

Jageshwar lying in the beautiful Jatganga valley, housing one of the 12 Jyotirlingas of India, is a complex of 12 temples. Two large fairs are held at Jageshwar, one on the occasion of *Shivratri* (February) and the other in the month of Shravan (July-August). A dip in the Jat Ganga and in the Brahmakund, near the temple complex, is of great religious significance on the occasion. Gananath Mela (Almora). Situated at an elevation of 2,116 meters above sea level, Gananath is 47 km from Almora. A fair is held here every year on Kartik Poornima (October/November), when thousands of pilgrims come to worship Lord Shiva and Ganesh. Childless couples also come here in large numbers to evoke blessings for the gift of a child.

Situated in the wide valley of Ram Ganga (West), Dwarahat is 38 km from Ranikhet and has a cluster of old temples. A huge fair is organised here every year on Baisakhi Day (April). Kasar Devi Mela (Almora) is held in the picturesque part of town twice a year during the summer and winter months. Swami Vivekananda came to Almora and meditated in a small cave at Kasar Devi.

Some festivals of Kumaon are unique and unusual. Through all fair and festivals have their own salient features, Nanda Devi, Rajjaat, Chhiplajaat, Hill Jatra, Kandali, Chaumu Jaat of Pancheshwar and Dahau (fish festival) of Ramganga valley have their unique way of cultural expression. The fair of Jauljibi, Thal and Bageshwar have been closely linked with the trading activities of people of this region, while the fair at Dwarahat, Syalde and Devidhura are important from the religious and cultural point of view.

The major fairs and festivals of the Garhwal region are Hatkalika Fair, Tapkeshwar Fair, Surkhanda Devi Mela, Kunjapuri Fair, Lakhawar Village Fair, and Mata Murti Ka Mela. The world famous Kumbh Mela/Ardh Kumbh Mela is held at Haridwar every twelfth/sixth year interval.

Arts and Crafts

The peace and tranquility of Uttarakhand laid the foundation for a treasure house of paintings and art. Out of the two major art forms, the art of stone carving and woodcarving are fairly well known. The art of stone carving gradually died down, but woodcarving continued. Woodcarving could be seen on almost every door of a Garhwali house until only half a century ago. Woodcarving can still be seen in hundreds of temple all over Garhwal. The remains of architectural work have been found at the Chandpur Fort, temple of Srinagar, Pandukeshwar (near Badrinath), Devi Madin (near Joshimath), and Devalgarh Temple.

Looking Towards the Future

Dev Bhumi Uttarakhand from where the holiest of rivers–the Ganga and the Yamuna–originate has always occupied a central position in the religious and cultural life of India. It is home to important pilgrim centers of the Shaivite, Vaishnavite and Shakta sects. The *Puranic* gods and goddesses of the greater tradition are worshipped in the form of various local deities throughout the region. The folk culture and society of central Himalaya culture and tradition could also be linked with the civilizations of Central and Southeast Asia. The comparative study of festivals and folk traditions from farming communities would not only restore broken historical links but also provide new insights from applying modern scientific viewpoint to traditional beliefs and practices.

The process of modernization has changed the social system and the great tradition of fairs and festivals. Thus the social, cultural and religious systems are under change. In the name of science, progress, poverty eradication, environmental awareness, fertilizers, medicine, rights of women and child, information technology, etc., they have marginalized and de-valued the rich knowledge systems, rich culture of fairs and festival etc. that local peoples possess.

One can see the incredible challenges ahead: to undo the damage created by modern systems and institutions, which do not acknowledge holistic learning spaces; to recover the lost value systems indigenous to our cultures, which have been steadily eroded to regenerate rural communities and our dying culture and preserving indigenous belief systems of *Dev Bhumi* Uttarakhand, the great tradition of fairs and festivals.

References

Agrawal, D.P. and Kharakwal, J.S. 1998. *Central Himalaya: An Archeological, Linguistic and Cultural Synthesis*. Aryan Books International, New Delhi, India

Atkinson E.T. 1884. *Himalayan Gazetteer*. North-west Provinces in Oudh Government Press, India.

Bhatt D. and Bhatt J.C. 2005. Harela: the farm festival of Uttarakhand. *Asian Agri-history*. 9 (3): 221-230.

Dabral, S.P. 2002. *Tehri Garhwal Rajya ka Itihas*. Vir Gatha Prakashan, Dogra, Garhwal.

Joshi, Anuradha. 2001. *The Child and the Family: A Study of the Impact of Family Structure upon Children in Rural Uttarakhand*. Mussoorrie: Sanshodhan.

Pant R C. 2002. A historical prospective of agricultureal interactions in Uttarakhand. *Asian Agri-history*, 6: 157-162.

Sharma, N. and B. Sharma. 1999. Children in difficult circumstances: familial correlates of advantage while at Risk. In T.S. Saraswathi (ed.) *Culture, Socialisation and Human Development*. New Delhi: Sage Publications.

Samvat Vikram and Berreman Gerald D. *Hindus of the Himalayas*. Oxford India.

Chapter 13

Traditional Knowledge System in Sustainable Eco-Development: An Uttarakhand Perspective

☆ *Deepak Pant, Anoop Singh and Piyush Malaviya*

Development, as a process of change, makes its impact both in space and time. Development dynamics is characterized by the pace and direction of betterment and is influenced by a wide spectrum of interactive factors including the natural resource potential, technological capacity, socio-cultural systems, economic forces and political motives. Human actions whether viewed as individualistic responses or responses of groups recognized by caste, religion or state, are basically the manifestations of human desire for good. There are no conflicts on this broad and ultimate goal of human race despite of enormous diversity in respect of level of development or concerns for development.

Economic conditions and policies have a profound effect on the ability of people to use biological resources sustainably, to conserve them, and to share their benefits. Some of the most effective tools for sustainable use and conservation of biodiversity, however, are strategies to strengthen social involvement and participation in biodiversity management. Without local participation and support, many conservation and sustainable use efforts fail. Many in the conservation and sustainable development communities have recognized this, and social components are now commonly found in eco-development projects. The use of traditional (indigenous knowledge) and strategies for local participation are the two important tools required for sustainable eco-development of any region.

People are ultimately the most important resource for designing and implementing of biodiversity management programme. This is recognized in the United Nations Development Programme's (UNDP) 'Capacity 21' programme designed in the wake of the UN Conference on Environment and Development. The programme is an effort to build the human capacity necessary to implement Agenda 21, and the international agreements signed in Rio de Janeiro (UN 1993). Indigenous people with a historical continuity of resource use practices often possess valuable knowledge about the behaviour of complex ecological systems in their own localities. This knowledge is accumulated through a series of observations that stretch over many generations, where indigenous peoples having depended on local environments for the provision of resources over long periods of time, often develop a stake in conserving biodiversity. Their practices for biodiversity conservation are built through lengthy trial and error processes. This implies an intimate relationship between the applications of ecological knowledge with belief systems.

Perspectives on Eco-Sustainability

The relation between environment and development has been widely discussed in recent years. This has usually been in the third world context and often in conjunction with the terms of population and resources.

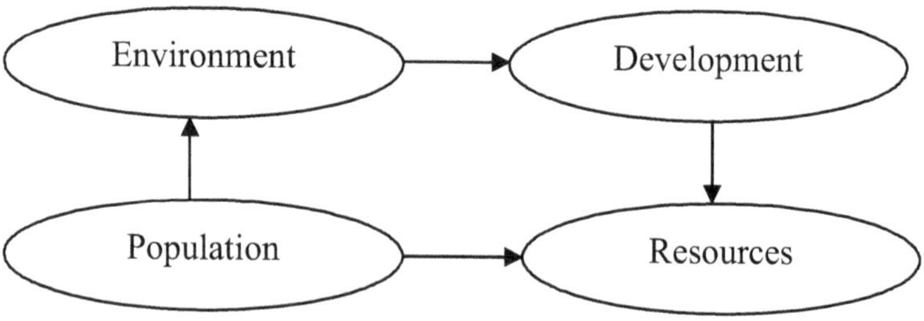

Sustainable development has been defined as 'development that meets the needs of the present without compromising the ability of the future generations to meet their needs' (WECD 1987).

Environment is a basic dimension of development especially in hills. Development must take into account the fact that mountain regions are ecologically most fragile of terrestrial systems. Soil conservation, land use, water resources, energy resources from both water and forests are all interdependent and of vital importance in integrated planning in hill areas. In context of sustainable development, several approaches have been used to explore the ecological part of the problem. Some key phrases are:

1. *Maximum sustainable yield (MSY) of a natural resource*: The implication is that a renewable resource such as forest or field crop may be safely harvested up to a certain level. As long as MSY is not exceeding the resource is sustainable.
2. *Carrying capacity of a region*: The implication is that a region can sustain a certain population of wildlife, farm animals, or people.
3. *Assimilative capacity of a region*: The implication is that it is acceptable to release pollutants into a watershed, air-shed or land-filled provided that environmental standards are not violated.

The need of formulating unifying approaches amenable to economical, social and environmental analysis has been stressed repeatedly following the emergence of sustainable development philosophy. Foy and Daly (1989) identified two distinct sets of approaches:

(*a*) Strong sustainable approaches where stress is needed complementary natural and manmade capital,

(*b*) Weak sustainability approaches where environmental degradation is permissible provided it is compensated through appropriate techno-scientific inputs.

Distinct from traditional educational processes, biodiversity management calls for considerable integration and cooperation among professions and academic disciplines in order to bring the best science, traditional knowledge and field know-how into practice. It also requires integration among public and private interests including business, communities, public natural resource managers, and individual farmers, foresters and fisher folk.

Experience indicates that one way to develop people's skills is to involve them in the preparation of national strategies and action plans enabling 'biodiversity planners' also to be 'biodiversity implementers', encourages policy makers, scientists, resource managers and local leaders to gain an understanding of the problems and opportunities in the field, and take responsibility for the implementation and evaluation of their decisions.

Indeed, on balance it appears reasonable to conclude that the many restraints on the use of natural resources that include protection everywhere to keystone resource species such as fig trees protection to highly susceptible stages such as birds breeding communally at a heronry, and protection to entire biological communities of sacred sites such as sacred groves or ponds, may have evolved culturally in response to the need to ensure more sustainable use of biological populations and conservation of biological diversity (Gadgil 1995).

In modern societies, as in traditional ones, protected areas serve the long-term interests of the members of society. The difference between them lies in the way in which nature is regarded, and in the trade-offs that are regarded as permissible. That is, the difference lies in the beliefs and traditions that determine how nature is regarded in each. Some, however, would argue that these differences might not be fundamental. Indeed Wilson (1984, 1993) believes that humans have an inherent inclination to affiliate with life and life-like processes and that these innate tendencies may be a basis for an appropriate ethic of care and conservation of nature, especially the diversity of life.

An innovative approach aimed at maximizing the skills of professional taxonomists involves the use of local rural adults as 'parataxonomists'-persons trained to collect, sort and prepare material for subsequent examination by taxonomists (Janzen, 1992; Janzen *et al.*, 1993)

Indigenous peoples without formal scientific training are generally dependent on their biodiversity to varying degrees and need to be able to recognize and communicate about species of value for food, medicine, clothing, construction, weaving, dyeing and other uses, as well as about those that are poisonous or harmful to humans, their crops, or their buildings. An immense amount of precision, as measured against modern species concepts, is to be found in some indigenous people's naming systems (Mayr, 1969).

The capture and utilization of this specialist ethnobiological knowledge, and its relation to the scientific database is an important aspect of biodiversity exploration. This is especially so as some tribal groups and their cultures are themselves dying out, and further because much relevant information is held now only by their more aged members as products of the modern age supersede the traditional at the village level. While significant progress has been made in relation to medicinal plants and land races of crops, only a fraction of that which exists has so far been documented. This is time-consuming work that requires personnel familiar with local languages and dialects who will be trusted by the indigenous peoples being visited. Nevertheless, it is appropriate to accord the capture of such indigenous knowledge priority (Wigstman *et al.* 1992; Smith *et al.* 1993).

Mountain areas occupy about one-fourth of the world's land surface and shelter about 10 per cent of the world's population. Another 40 per cent population occupies the watershed areas below mountains. Thus half of the world's people depend directly or indirectly on mountains. Mountains provide water, energy, minerals, forest and agricultural products and are areas of recreation. They also store the biological diversity necessary for the sustainability of human life. Many varieties of plants and animals have originated in the mountains. Mountain people cultivate many lesser-known crops and medicinal plants. This biodiversity is of immense value for the future, for genetics and breeding. The mountains also provide scenic beauty and recreation. Mountain areas are considered to be the abode of Gods and Goddesses and have an important place in religion and mythology (Gangwar, 2002).

In the mountain ecosystem, forest plays a crucial role in maintaining the delicate balance between the climate and overall temperature of the region. With increasing deforestation and rampant loss of tree cover; a subtle imbalance in the climate and temperature is being created. The snow capped mountains and ice covered glaciers, which in earlier times remained snow bound started to melt at greater speed than before. This released higher volumes of water in the mountain rivers leading to wide spread flooding along river courses. Accompanied with high density of rainfall in the monsoons, much of the top soil would get washed away due to the absence of any forest cover, thus further affecting the fertility of the soils (Chadha, 1997).

Role of Traditional Knowledge System (TKS) in Sustainable Development: An Uttarakhand Perspective

The Himalaya, lying in Indian Territory, is spread over a length of about 2,500 km and a width of 220 to 300 km. It covers partially eleven provinces of India *viz.*, Jammu and Kashmir, Himachal Pradesh, Sikkim, Nagaland, Manipur, Mizoram, Tripura, Meghalaya, Assam, West Bengal and Uttarakhand. It has a total geographical area of approximately 591 thousand km^2 inhabited by 51 million persons. The region is characterized by sparse population, undulating terrain, far flung small villages difficult to approach, scattered land holdings, shallow and gravely soil, agro-pastoral economy, scanty irrigation, lack of technological advancements. Abundance of natural resources and their diversity offer positive values from the point of economic growth as well as environmental

Figure 13.1

conservation. Coexisting with these positive values are negative values. Low level of advanced technology, financial capacity, market network and connectivity account for poor entrepreneurship in indigenous population (Rao and Saxena, 1994).

Uttarakhand, formerly a part of Uttar Pradesh, consists mostly of hilly areas. Mountain areas in the past were ignored in respect of evolving scientific understanding of natural resource dynamics and evolutionary and organizational development of human system. Various reasons accounting for the negligence of mountain ecosystems include orthodox views about mountains associated with catastrophes and poor connectivity. Patterns and processes influencing natural as well as human systems in the mountains influence not only the mountain people but also those living in adjacent low land areas. Development prospects of 90 per cent of world population living in 75 per cent of earth's low land surface is initially found to what happens or does not happen in high lands (> 1000 m altitude) which occupy only 25 per cent of world population (Eckholm, 1975).

Himalayan region previously covered in the state of U.P. consists of two sub regions *viz.* Garhwal and Kumaon, was carved out on 9 November, 2000 to form the state of Uttarakhand. The state of Uttarakhand encompass an area of 53,485 sq. km., which accounts for nearly 15.5 per cent of the total geographical area of western Himalayas and 1.63 per cent of the total land area of India. It lies between 28′44" N to 31′28" N latitude and 77′35" E to 81′01" E longitude. The total population of the state is 84, 79, 562 with a male: female ratio of 1000:964 (Census: 2001). The state comprises of 13 districts namely Chamoli, Pauri, Tehri, Uttarkashi, Dehradun, Hardwar and Rudraprayag in Garhwal region and Nainital, Almora, Pithoragarh, Udham Singh Nagar, Champawat and Bageshwar in Kumaon region.

The state can be divided into three physiographic regions *viz.*, the Himalayas, the Shivaliks and the plains. The state has a temperate climate except for the plain areas that have a subtropical type of climate. There are six forest types known to occur in the state, which include tropical moist deciduous forests, tropical dry deciduous forests, sub-tropical pine forests, Himalayan moist temperate forests, Himalayan dry temperate forests, sub alpine and alpine forests (Champion and Seth, 1968). The total area of the forests in

Uttarakhand is 34661.52 sq. km., which works out to 64.81 per cent of the state's geographical area.

The fast deterioration of climatic and socio-ecological balance in this region has become a matter of deep concern. Due to illegal and excessive exploitation, populations of some of the medicinal plant species such as *Aconitum heterophyllum*, *Podophyllum hexandrum*, *Orchis latifolia*, *Nardostachys grandiflora*, *Taxus baccata*, etc. have now become rare, endangered and threatened (Misra, 2003).

Crop diversity has eroded very rapidly in the Himalayan region. This was reported by Maikhuri *et.al.* (1997). Traditionally, over 40 crop species and hundreds of varieties of cereals, millets, pseudocereals, pulses, oil seeds, tubers, bulbs, and spices have been cultivated in these hills. However, in recent times this has changed, with increasing homogenization coming into the agricultural practices. The area under traditional crops in the Alaknanda valley, for instance, has declined substantially in the last two decades: in the case of oat (*Avena sativa)*, buckwheat (*Fagopyrum* spp.), naked barley (*Hordeum himalayens*) and legumes such as cowpea, matbean, adjuki bean (*Vigna* spp.) and horsegram (*Macrotyloma uniflorum*), the decline was of the order of 72-95 per cent. These have been replaced by cash crops (potato, soyabean, kidney bean, mustard, amaranths). 65 per cent of the area under hog millet (*Panicum miliaceum*) and foxtail millet (*Setaria italica*) is now used for high-yielding rice varieties and soyabean. Bhangjeera (*Perilla frutescense*), horsegram (*Macrotyloma uniflorum*), and legumes of *Vigna* spp. are almost extinct now. Several traditional varieties of rice (The British official Atkinson reported in his Himalayan Gazetteer of 1882 that there were 48 distinct varieties and thousands of nondescript varieties) are now hardly grown, having been replaced by a handful of High Yielding Varieties (HYV).

The care for nature and the zeal to protect nature's resources has been deeply enshrined in the Uttarakhandi's psyche. This is best exemplified in the famous 'Chipko' (Hug the tree) movement of the 1970s (Bahuguna, 1981) and the 'Maiti' movement in recent times. These are not small movements in isolation but caught people's fancy and they embraced the message by and large. Thus community participation, which is a must for any grass root level movement to succeed, has been successful in Uttarakhand. In 1982, commercial logging of all live trees was banned for fifteen years in the eight hill districts of Uttar Pradesh in Himalayas.

Another example of people's participation in eco-development of Uttarakhand (in the form of forest regeneration) was made by the residents of the village Jardhargaon nestled in the hills of Hemval Ghati, one of the numerous picturesque valleys of the Tehri Garhwal Himalaya. Over a decade ago, the hills around the village had been almost completely denuded, the result of felling both by contractors as also the villagers themselves. Shortage of fuel and fodder, soil erosion, and a host of other problems became daily occurrences. In 1980, the Gram Sabha (village committee) decided that something had to be done to re-green the area. Inspired and helped by the Chipko Movement, they formed a Van Suraksha Samiti (forest protection committee), consisting of village youths who were paid in cash or kind by every family of the village. This samiti ensured that a large area of hillside above the village was left totally alone, with no grazing or fuelwood collection, and no felling by contractors or the Forest Department. A bit of tree planting was done of native species like oak and walnut. However, most of the area was simply left to regenerate on its own, since the villagers had faith in the power of nature to recoup. No expensive fencing was done, simply a social sanction against use of the area, and the dedicated efforts of the committee members. A dozen years after protection started, the results have been astounding. A patch of several hundred hectares of mixed forest now covers much of the slope above the village. This forest is rich in biodiversity and portrayed oak (*Quercus leucotrichophora*), burans (*Rhododendron arboreum*), horse chestnut (*Aesculus indica*), walnut (*Juglans regia*), bimal (*Grewia oppositifolia*), chir pine (*Pinus roxburghii*), bamboo, and other species along with small wildlife (boars, deers, bears etc.) present in a wild profusion. In the last few years, the villagers have also developed a couple of nurseries, and are using it to afforest other parts of the barren slopes around Jardhar. The Van Suraksha Samiti is looking after these nurseries with help from the Mahila Mangal Dal (the village women's committee).

Even more remarkable than this forest regeneration and protection effort are the attempts of one farmer and social activist Vijay Jardhari, resident of the same village Jardhargaon, in reviving and modifying traditional crops and cropping methods. Despite the fact that most farmers have switched to high-yielding varities of seeds and new techniques of farming, Vijay and some other farmers still maintain plots of what they call *baranaja*. Literally meaning twelve grains, this age-old practice involved sowing of a mixture of crop seeds into a single plot of land. *Rajma* (Kidney beans), *urad* (Black gram), *mung* (Green gram), *kulth* (Horse gram), *marsha* (Amaranthus), *jhangora* (Barnyard millet), *mandua* (Finger millet), and other crops were grown in a jumbled profusion, which to a modern agriculturist would look like a big mess, but which constituted an extremely well thought out way of getting sustainable yields. Since maturity periods of these crops varied, different crops would be harvested at different times. Leguminous plants like pulses would keep up soil fertility. The result was higher and diverse yields which met a diversity of needs. Along both the *baranaja* and other fields, fruit trees and other plants would be grown, to provide additional food, fibre, fuelwood, and other products of daily need. Vijay has not stopped at just trying a diversity of varieties in his own field. For the last few years, he has been going around the Tehri Garhwal region, collecting indigenous crop varieties, and documenting their characteristics as told to him by local farmers. He has also made a herbarium of rice and other varieties, stalks and grain heads carefully pressed, characteristics neatly written down. His collection includes *Thapachini* (tall, high-yielding, non-lodging), *Nagni basmati* (red and white grains, nice aroma), *Gorakhpuri* (fast-growing), *Chowari* (red grains, high-altitude tolerant, starchy), *Bangooi* (black-stemmed), *Noori* (early maturing) and so on. All these efforts of Vijay Jardhari laid the foundation of the 'Beej Bachao Andolan' (BBA), which began in the late 1980s, and is now more than fifteen years old. The 'Beej Bachao Andolan' (Save the Seeds Movement) is not only a crusade to conserve traditional seeds but also to promote agricultural biodiversity, sustainable agriculture and local traditions. The hallmark of the BBA is that it is a people's campaign and flourishing without any government financial assistance or help. However, with the provisions in the Trade-Related Intellectual Property Rights (TRIPs) part of GATT coming into force, especially those seeking to harmonise IPR regimes across the globe and to enforce patentability of life forms, are forcing changes in the Indian Patent Act and related legislation. This could have severe implications for biodiversity and farmers' rights. IPRs are expensive, and corporations would try to push their protected seeds over as wide an area as possible to recover costs and make profits. Further displacement of traditional diversity and homogenisation would result. Additionally, innovations by farmers, which result in expanding diversity, may be hindered if IPR regimes favour the formal sector breeder at the cost of the farmer (Kothari, 1999).

On a similar note, the civil soyam forests in many parts of Uttarakhand managed by the local communities also give an idea about the people's participation in sustainable eco-development (Negi 2003). The Van Panchayat system, wherein local villagers are involved in the protection and management of surrounding forests, with overall coordination and control vesting in the local administration, is another example of people's involvement in sustainable development of Uttarakhand.

Traditional Knowledge is the total stock of knowledge and practices developed and used by a community over centuries to meet different life situations and include traditional medicine and healing practices, folklore and folk art, ethnic and anthropogenic products. Uttarakhand by virtue of its unique geographical location and rich natural heritage has an extensive tradition of folk art and craft, folk music, folk dance and biodiversity as well. Historically also, Uttarakhand has been the abode of a large number of saints and sages for centuries, which during the course of their living developed the science of Ayurveda.

In the traditional communities, the exploitation of nature and its resources for personal greed is largely unheard of. Most of these communities are living in perfect harmony in nature since times immemorial. However, with the social environment gradually changing, it is upon the newer generations to adapt to these changes and add to the existing stock of knowledge. Mountain people of Uttarakhand represent thousands of years of accumulated experience living and working in their rugged and awe-inspiring environments, and are intimately connected to the natural environment that sustains them. These traditional cultures are themselves fascinating and sources of great environmental wisdom.

In recent times the burden on traditional knowledge system (TKS) has increased tremendously. Conservation of natural resources of global significance can be ensured through economic incentives and livelihood options for local people, who are dependent on resource use. Consequently, the management of Indian Biosphere Reserves has given due consideration to alternative livelihood options for dependent populations, and other methods for socio-economic upliftment of the people. Other strategies adopted includes: demarcation of user zones, strengthening the legal framework, use of economic instruments, compensation and rehabilitation packages, maintenance of corridor areas, development of communication systems and networking, and development of nondestructive practices such as eco-tourism. This will enable local people to attain a balanced relationship with the environment and to participate in planning and management of the protected area. The existing forms of development, however, have introduced a process of devaluation and marginalization of indigenous knowledge and skills, which in turn likely to lead to gradual eclipse of this knowledge.

Development and erosion of knowledge systems is subject to social, economic and political forces, which facilitate the consolidation, circulation and transmission of particular forms of knowledge while eroding others. The introduction of the mainstream development programs in the mountain communities has reportedly resulted not only in the loss of land, but more significantly in the decimation of their unique knowledge, culture and identity and in the decline of women's social position. However, the realization of an ecological crisis and the problem of sustainability have brought about limitations of the systems of knowledge on which the present development has been based. This has had a number of consequences, both with regard to the sources of knowledge and the methods of enquiry.

Uttarakhand has declared itself as an herbal state and plans to develop medicinal plant sector as a priority area. Also, keeping up with recent trends it has gone completely organic and declared itself as an 'organic state'. One hopes that these declarations will bear fruit in the long run in not only protecting the diverse flora of the state but also protect it from the harmful effects of agricultural chemicals. Despite the fact that Uttarakhand state has realized the importance of medicinal plants as a critical and valuable resource, the state has still a long way to travel. The state has a good network of research and developmental organizations in the government and non-government sectors to facilitate the process of medicinal plants conservation. The establishment of State Medicinal Plants

Board is expected to provide the much needed guidance and direction to the sector in the state.

Eco-Sociological Changes

The isolation and impediments immobility favored enormous diversity in socio-cultural systems adapted to local conditions and evolution of rich traditional knowledge on resource dynamics, use and management. Hill societies in historical times have laid emphasis on self-reliance to secure their perpetual existence in isolated and inaccessible areas. However, in recent times increasing pressure of population have meant greater need of hill people for food, fuel, fodder and fibre all of which the limiting as well as depleting resource base could not provide. This depletion is aggravated by rising demand from plains for forest like timber, fuel, charcoal, mineral, gums, resins and medicinal plants. Tourism is increasing with the result that the pressure on land and infrastructure has also increased. Massive increases of tourist activity in fragile mountain ecosystems pose a serious challenge in the developed world, and an even more daunting one in developing nations.

Mountain communities are amongst those marginalized groups whose destinies are mostly decided by an influential tribe of resourceful institutions/persons who often represent diverse interest groups. Be it in the name of conservation or tourism, the natives always find them as the last party to decide their own destinies. Perhaps it is because they live in a land of abundant resources and opportunities. Perhaps, it is because of their looking at the mountains from an all-together different perspective. Or it may be due to their isolation or poverty or a myriad of other esoteric reasons. However, local communities have now started asserting themselves as the rightful owners of these lands and also started demanding their rights as stakeholders in whatever form of development these regions are going through, be it tourism or timber (Rana *et al.*, 2003).

Grazing and agriculture are the most primitive and the most natural occupations of the man of Uttarakhand. These perfectly suit his temperament and if carried out within permissible limits blend with the nature and ecosystems harmoniously. But the ecological balance is upset when the resources of nature are taxed beyond limit. In Uttarakhand, cultivation of steep, often unterraced hill slopes can be found as a rule rather than exception. This practice affects the whole nation by flooding of the rivers, choking of the irrigation channels, silting of the reservoirs, bursting of the dams and premature dysfunction of hydroelectric power projects. Similar has been the harmful effect of indiscriminate grazing on the pasturelands, which have been badly impoverished. Agriculture has been extended at the cost of forests which have survived, outside the boundaries of the reserve forests, only at such places which were unfit for agriculture due to rocky and rugged terrain. Centuries of unabated grazing had almost destroyed the forests, and extension of agriculture has completed this circle.

Despite of all the rhetoric of joint management of forests and despite their very strong dependence on forest food resources, women and men have very little control over the way forest resources are to be used (Khan, 2005).

Root Cause of Environmental Changes

Several factors are responsible for above-mentioned changes, major among which are deforestation, construction activities and tourism. Deforestation means felling of forest trees without regard to their vital role in the life support system and ecological balance. The construction of new roads and widening of existing ones has promoted deforestation, landslides and erosion. During road construction huge portions of fragile mountainous region are cut and thrown into adjacent valleys and streams, which weakens the already weak mountain slope and increase the turbidity of the streams. As a result of the dam construction activities, the geomorphology of the Bhilangana and Bhagirathi rivers is altered to a great extent. The tunneling material and byproducts of explosives are disturbing the natural setup of the Bhagirathi near dam site (Sharma, 1980). Besides the dam is in earthquake prone area so the future risks of safety are great.

In Uttarakhand, deforestation is not a recent phenomenon. Degradation of forest cover in Uttarakhand has given way to many other problems such as soil erosion, slope failures, depletion of soil fertility, scarcity of fuel wood and fodder, increased burden over land flows, reduced ground water recharge and loss of biological diversity. This deforestation could be responsible for changes in Uttarakhand climate but the state of knowledge is confusing and it is difficult to establish a reliable forest-climate relationship. More drastic damage and destruction has occurred in the past three decades than ever before because of haphazard, undisciplined and miscalled 'developmental activities' and free permission given to those guilty of a most ruthless assault on natural resources in pursuance of shortsighted economic policies.

Excessive erosion is produced by exceptional meteorological events mostly after man has destroyed or diminished the protective vegetative cover of the slopes or changed the run-off pattern of slopes and rivers with roads, canals or terraces. Oversteep road cuts reducing the stability of slopes are another factor contributing to excessive erosion. The greater the harmful human interference, the smaller a meteorological event is needed to cause or perpetuate excessive erosion.

Modern tourism demands more comforts and so it exerts more pressure on already exploited natural resources. In fact, the promotion of tourism in fragile mountainous regions should be done very carefully so that the landscape may be maintained intact and the evils of tourism may be minimized. According to Singh and Kaur (1985) tourism may be effectively planned for the good of the guests and the hosts, it is necessary for the region to know what it has, what it can sell and what it should not, and if possible how to sell and whom to sell. Resources are fragile, too vulnerable or too precious to be sold.

Regional development perspective of Uttarakhand seems to be clouded with environmental and economic issues; social issues being usually given a peripheral or secondary importance. Environmental degradation is more a problem of relationship among people competing for productive resources than of relationships between people and habitats (Horwitz, 1988). Development priorities for the mountains are justified for improving upon the life quality of environmental people and also to persuade the people of the plains that the future of the mountains cannot be separated from their own (Eckholm, 1975).

Mountain societies have experienced widening socio-economic disparities and deepening gender inequalities. Three major constraints in the development of mountain peoples relate to (*a*) interventions from outside, which have by and large been extractive, (*b*) the mountain peoples' own fragile production structures, further threatened by these extractive external relations, and (*c*) weakening institutional mechanisms.

A Framework of Sustainable Eco-Development in Uttarakhand

Mountains offer both opportunities and constraints for development. Mountain societies adapt to these constraints and opportunities through trials and errors. Development interventions involving technological and institutional changes attempt to increase the pace of development. Development policies and programmes have consistently failed to give careful attention to mountain areas. Even if it is done, approaches are organized and managed in an inappropriate manner resulting in many adverse impacts, particularly on the environment.

Development strategy is basically a set of actions identified by choices and compromises in decision-making process. Development concerns in Uttarakhand revolve around how resources of the region could be managed for conserving/improving the environmental values of the region together with socio-economic development of mountain people. While conservation of natural resources figures top priority on agenda of environmental activists, thinking about the development of mankind in wider perspective, ways of building upon the economical potential linked to infrastructure development, introduction of advanced technology and increased cash flow through market economy have also gained ground. These incidentally are the primary concerns of deprived and desperate rural hill men. The native people of the Himalayas perceive developmental process from holistic point of view. In such a paradigm, the development of the individual as well as the community is important.

During the last two decades, the link between biodiversity conservation and sustainable socio-economic development has been recognized. It appears that the interests of some human groups have been strongly linked to the prudent use of their resource base, and that they have evolved appropriate conservation practices based on some simple and approximate rules that have tended to ensure the long-term sustainability of the resource base. A process of trial and error may have developed these rules, with acceptance of practices that appeared to keep the resource base secure coupled with rejection of those practices that appeared to destroy the resource base.

Strategy Planning for Sustainable Eco-Development

The logical unifying principles of inducing sustainable development in Uttarakhand as outlined by Rao and Saxena (1994) could be:

1. Realize the opportunities and constrains for development of mountain perspective.
2. Divert attention of people from 'what they do not have' to 'what they have' for overcoming the present state of desperation.
3. Build on the local knowledge, natural resources, skill and human capacities.
4. Proved people with technology and materials altogether new to them or beyond their reaches but meeting sustainability criteria.
5. Make people to realize the real cost of given interventions by way of promoting their participation in development.
6. Design people-government participatory action frame works clarifying responsibilities, accountability and profit/loss sharing.
7. Properly managed ecotourism has the potential to generate revenue for communities through conservation-linked enterprise development.

Conclusions

Faiths, cultures and traditions give people their basic orientations toward the natural world and guide their actions. Nature has been considered (and is still considered) by many people as an obstacle to human purposes as well as the direct or indirect source of all the material necessities and comforts of human life. For instance, aquatic ecosystems are considered both as a reservoir of biological resources (fish, shrimps) to be preserved, and the reservoir of diseases (malaria, dengue etc.) to be eradicated. This dilemma may explain why human attitudes to nature differ from culture to culture, and have changed over time. This importance is often overlooked in conservation programmes, while people's commitment to conserving and sustainable development using biodiversity springs from the human 'sense of place'. There is a need to understand better, how ethical norms and religions condition human behavior toward biodiversity.

There is a need to recognize the value of traditional knowledge, and subsequently to develop a mechanism for the appropriate protection of, and compensation for, such knowledge (UNEP 1994). This can be achieved through (1) a better knowledge of biological resources being exploited, as well as the full range of uses and values of these resources, and (2) compilation of available information with the support of a group of specialists. We must also identify and develop means to maintain traditional knowledge and to strengthen and develop indigenous and local community strategies for conservation and sustainable use of biodiversity, fully respecting their intellectual and cultural integrity and their own vision of development. This research should build a greater understanding of the relationship between biodiversity and local systems of knowledge and resource use, and should translate this understanding into useful policy (USNRC, 1992).

We therefore need to conserve the diversity of traditional resource management practices and systems if we want to construct a new resource management science better adapted to the real world. The rejection of the monolithic scientific vision of resource management does not mean the overall rejection of science (Gadgil and Berkes, 1991). The task is to develop a flexible approach by conserving what is useful in science. Ecology is in a unique position to be the cornerstone of a new science of resource management that synthesizes the best of the old and new wisdom towards a more sustainable future. But ecology would first have to extricate itself from the older utilitarian, 'control over nature' tradition of resource management.

Note: Views expressed in this article are of author's own and do not necessarily represent the views of their organizations.

References

Bahuguna, S. L. 1981. *Chipko: A Novel Movement for Establishment of Cordial Relationship between Man and Nature*. Chipko Information Centre, Silyara, Tehri. pp. 1-13.

Chadha, S. 1997. A people's movement for Indian high Himalayan watershed management-beyond chipko movement. PWMTA farm field document no. 7 *In: Participatory process for integrated watershed management.* P. N. Sharma (ed.). Participatory watershed management training in Asia (PWMTA), GCP/RAS/161/NET-RAS/93/062, FAO (UN), Kathmandu, Nepal. pp. 68-71.

Champion, H. G., and S. K. Seth. 1968. *A revised survey of the forest types of India.* Government of India Press.

Eckholm, E. 1975. The deterioration of mountain environments. Science. 139: 764-770.

Foy G, Daly H. 1989. Allocation, distribution and scale as determinants of environmental degradation: case studies of Haiti, El Salvador and Costa Rica. Washington D C: World Bank.

Gadgil, M. (1995). Traditional conservation practices. In: *Encyclopaedia of Environmental Biology.* Academic Press, New York.

Gadgil, M. and Berkes, F.1991. Traditional resource management systems. *Resource Management and Optimization.* 18:127-141.

Gangwar, Abdhesh. 2002. Hardy Mountains, Fragile Environments. Workshop on "Conservation of Heritage and Environment of Hill Stations and Hill Areas". Centre for Environment Education, Himalaya Initiative.

Horowitz, M. M. 1988. Anthropology and the new development agenda. Development Anthropology Network. 6: 1-14.

Janzen, D.H., Hallwachs, W., Jimenez, J.and Gamez, R. 1993. The role of the parataxonomists, inventory managers, and taxonomists in Costa Rica's national biodiversity inventory. In: Reid, W.V., Laird, S.A., Meyer, C.A., Gamez, R., Sittenfeld, A., Janzen, D.H., Gollin, M.A. and Juma, C. (eds.), *Biodiversity Prospecting: Using genetic resources for sustainable development.*223-254. World Resources Institute, Washington DC.

Janzen, D.H.1992. A south-north perspective on science in the management, use, and economic development of biodiversity. In: O.T.Sandlund, K.Hindar, and A.H.D. Brown, (eds.),

Conservation of Biodiversity for Sustainable Development, 27-52. Scandinavian University Press, Oslo.

Khan, T. I.2005. Biodiversity: Concepts and need of conservation. In: T.I.Khan and Y.S. Shishodia, (eds.), *Biodiversity Conservation and Sustainable Development*, 124-139. Pointer Publishers, Jaipur.

Kothari, A. 1999. Agro-Biodiversity: The Future of India's Agriculture. Article for MCAER Book; 7 February, 1999.

Mayr, E. 1969. The biological meaning of species. *Biological Journal of the Linnean Society*. 1:311-320.

Maikhuri, R.K., Semwal, R.L., Rao, K.S., Nautiyal, S. and Saxena, K.G. 1997. Eroding Traditional Crop Diversity Imperils the Sustainability of Agricultural Systems in Central Himalaya. *Current Science* Vol. 73 No. 9, 10 November, 1997.

Misra, M. K. 2003. Base Line Information on Medicinal Plants Conservation and Sustainable Utilisation: Uttarakhand. UNDP/GEF, MoEF, Government of India and Coordinated by Foundation for Revitalisation of Local Health Traditions (FRLHT) Bangalore.

Negi, S. S. 2003. Land, Water and Environmental Management. In: *Natural Resource Management in the Himalaya*. A.P.H. Publishing Corporation, New Delhi.

Rana, Dhan Singh, Sunil Kainthola and Pratibha S. Naithani. 2003. The struggle for community based conservation and equitable tourism in the Nanda Devi Biosphere Reserve in India. Paper presented at The Namche Conference: 24-26 May, 2003 "People, Parks and Mountain Ecotourism."

Rao, K. S. and Saxena, K. G. 1994. Sustainable development and rehabilitation of degraded village lands in Himalayas. HIMVIKAS publication no. 8. Bishen Singh, Mahendra Pal Singh. Dehradun, India.

Sharma, R. C. and Singh, H. R. 1980. Impact of Bhagirathi blockade on its riverine ecosystem. Johsard. 4: 35-37.

Singh, T. V. and Kaur, J. 1985. In search of holistic tourism for the Himalaya. In: Integrated Mountain Development. Eds. T. V. Singh and J. Kaur. Himalayan Books, New Delhi. pp. 365-401.

Smith, N., Wididburu, B., Harrington, R. N. and Wightman, G.1993. Aboriginal plant use from the Victoria river area of Northern Australia. *Northern Territory Botanical Bulletin*. 16: 1-64.

UN.1993. *Agenda 21, Rio Declaration, Forest Principles: The Final Text of Agreements*. United Nations, New York.

UNEP.1994. Report of the open-ended intergovernmental meeting of scientific experts on biological diversity. UNEP/CBD/IC/2/11.

USNRC. 1992.Conserving Biodiversity. A research agenda for development agencies. National Academy Press, Washington, DC.

World Commission on Environment and Development (WECD). 1987. Our Common Future. New York: Oxford University Press.

Wightman, G., Roberts, J.G. and Williams, L. 1992. Aboriginal plant use from the Elsey area, northern Australia. *Northern Territory Botanical Bulletin*. 15: 1-59.

Wilson, E.O. 1984. *Biophilia: The human bond with other species*. Harvard University Press, Cambridge, Mass.

Wilson, E.O. 1993. Biophilia and the conservation ethic. In: Kellert, S.T. and Wilson, E.O. (eds.), *The Biophilia Hypothesis*. Island Press, Washington, DC. and Shearwater Books, Covelo, California.

Chapter 14

Indigenous Knowledge Systems: Concept, Problems and Rationales for Studying IKS of Mountain Regions

☆ *O.P. Mishra*

Introduction

Twenty per cent world's land surface area is occupied by mountains. In India, mountain cover 2500 km long range and 236000 km ^{2} area in two flanks–the North western Himalayas comprising of states Jammu and Kashmir, Himachal Pradesh and Uttarakhand, and the North Eastern Himalaya including seven north eastern states (Assam, Arunachal Pradesh, Manipur, Mizoram, Meghalaya, Nagaland and Tripura) Sikkim and Darjeeling district of West Bengal (Melkania, 2002).

Mountains have always been evoking our vision, our responsibilities towards the vital life-nurturing parts of the nature. Ecological services of the mountains are crucial and of global importance. Mountains' importance, however, was undermined in the recent past and they were regarded only as the hinterlands of the mainstream areas. It is why they are in the throes of vicious changes and socio-economic and ecological cries. Cries of the mountains do not remain confined only to a small area or the region; they have widespread repercussions even on larger economies of the plains and on people in the downstream areas (Gautam, 2004).

Uttarakhand, the new Indian state in the Himalayan Mountains, has to meet many challenges in the way of its holistic development. This mountainous state is located almost in the midst of the Hindu Kush-Himalayan Range and has everything that could be the best in the whole region. Its biodiversity and indigenous wisdom is unique.

The low level of progress so far made on achieving sustainable development in mountain areas of India, suggests that Governments/policy and decision makers, researchers professionals, development specialists and donors have not sufficiently understood rural resource users or indigenous people and their knowledge in mountain areas. They have not understood these peoples' priorities, ways of making natural resources utilization and management decisions.

Problems and failures encountered in fostering development in communities residing in mountain areas have during recent years prompted some considerations in the study and understanding of indigenous knowledge systems. Van Vlaenderen (1999) points out that development agencies, theorists and academicians have come to realize that development efforts that are based on conceptualizations, methods and values of the Euro-American World often fail to bring about appropriate and sustainable development in the communities which they target. In analyzing these failures, it has become obvious that indigenous people's knowledge/practices, needs, values, priorities and capacities form an essential base for effective development programmes.

The Concept of Indigenous People and their Knowledge

Although the notion and interest on indigenous people and their knowledge/practices are recent, there are already as many definitions of indigenous knowledge systems and practices as there are writers on the subject.

According to the Inter-Commission Task Force (UNEP, 1992), indigenous knowledge may be defined as local community based systems of knowledge, which are unique to a given culture or society and have developed as that culture has evolved over many generations of inhabiting particular ecosystem. It also includes innovations and practices indigenous and local communities embodying traditional lifestyle relevant for the conservation and sustainable use of biological diversity. Indigenous knowledge system encompasses knowledge commonly known within a community or people, as well as knowledge that may be known to the tribal elders, a lineage, group, or a gender group (IUCN, 1997). Indigenous people are those people who originate from a particular community, locality or country and who have lived in that area for time immemorial.

Problems associated with indigenous people and indigenous knowledge systems

There are many problems associated with indigenous people and indigenous knowledge system. Some important problems are given below:

1. Inadequate publication and storage methods of indigenous knowledge
2. Secrecy and unwillingness to pass on knowledge to others
3. Lack of appreciation, commitment, recognition and willingness to promote indigenous knowledge systems/practices

Problems associated with indigenous people and indigenous knowledge has had far-reaching or serious impacts on development endeavours. Growing bodies of evidence indicate that one of the main reasons why "conventional" development approaches have failed is that they have tended to ignore the indigenous knowledge systems (Morin-Labatut *et al.* 1992; Salas 1994; and Warren 1991). This has led many people to argue that in order to ensure a more socially and ecologically sound approach to development, it is necessary to understand, respect, value and utilize the indigenous knowledge systems (Berkes *et al.* 1994 and Hech and Pscy 1989).

It is extremely difficult to acquire indigenous knowledge which is regarded as secret or confidential. Moreover, since indigenous knowledge is not documented but stored in peoples' minds, it bears a high risk of losing it. Maundu (1995) states that indigenous knowledge is diminishing at an alarming rate with the ageing of those in the indigenous population with strong links to the past.

Existing Initiatives to Address the Problems

Concerted and deliberate efforts are now being made by researchers/academicians to conduct studies on indigenous knowledge and practices in order to understand them and apply the same for sustainable development. This emerging interest will facilitate the achievement of sustainable development.

A number of African countries have undertaken various initiatives to integrate indigenous knowledge in agriculture, health care, environmental conservation and education programmes. The World Bank is supporting several of these initiatives. Few examples include:

Tanzania

The World Bank in partnership with FAO sponsored a national indigenous knowledge workshop in June, 2001. During the same year, the office of the Vice-President convened a consultative forum to develop a national indigenous knowledge strategy and action plan. Similar efforts are also underway to build local capacity of partners in Kenya Mozambique and Ghana.

Uganda

The World Bank sponsored a national workshop that resulted in the 1999 Kampala Declaration on indigenous knowledge for sustainable development. The Bank also provided a grant from the Institutional Development Fund to support activities to help integrate indigenous knowledge into the operations of health and agriculture ministries and build local capacity to document and exchange indigenous knowledge in Uganda.

Ghana

The Northern Savana Biodiversity Conservation Project seeks to improve the environment, livelihood and health of communities through sustainable natural resource management. A recent enactment by Parliament of a Traditional Medicine Practices Bill recognizes the important role traditional knowledge plays in providing primary health care.

Eritrea

The Integrated Early Childhood Development Project seeks to improve child health, child and maternal nutrition, early childhood education and care. The project includes a framework for identification, validation, collection, storage and dissemination of indigenous knowledge related to early childhood development and the design of a database of such practices to facilitate their exchange among local communities and across the region.

India

Importance of learning from farmers has been long realized. Honey bee network of Dr AK Gupta and Asian Agri-History Foundation of Prof. YL Nene are doing very commendable work in this direction. Some works have also been done under National Agricultural Technology Project on this aspect. Less effort has been made to document and validate IK of Uttarakhand livestock farmers.

Ways and Means of Achieving Sustainable Mountain Development through the Use of Indigenous Knowledge

Government should provide Policy Support to Indigenous Knowledge (IK). Changes in Educational Systems are required. Efforts should be made to include IK in course curriculum. Re-orientation of Research towards IK will also help in this direction. Establishment of Indigenous Knowledge Resource Centers for technology backup and support will be a good step in this direction. Identification of Indigenous Knowledge Systems/Practices and their Custodians is must for achieving sustainable development of mountain area.

Rationales for Studying IKS of Mountain Regions

Mountain ecosystems sustain an estimated ten per cent of the world population. Even people living in the plains/low land areas depend on the mountain ecosystems for resources like, water, energy, food etc. Mountain people are fully dependent on the natural resources, livestock and traditional agriculture. The farmers' indigenous knowledge is the product of centuries of trials and errors, natural selection, observation that can form a knowledge base on which researchers and extension workers can plan their research strategies and experimental procedures. Wang (1988) felt that through documentation of this knowledge, the indigenous capabilities will be strengthened and IKS could be saved from being pushed in oblivion, a good practices confined to one region may be transferred to other regions to solve the similar problems. Also a simple practice can be improved slightly to replace an innovation which finds difficulty to take root into society. Reasons for documentation of indigenous knowledge could be: (1) to understand scientific rationale, (2) to accelerate technological change, (3) to enable better understanding and technology development and newer concepts, (4) to increase awareness among the younger generation and to revive and restore pride among the farmers themselves.

While spelling out practical worth of such studies, some scholars made a plea for studying traditional practices which promise a number of useful contributions. First, through Extension Programmes, we are trying to change the technologies hitherto in use by our farmers. But the degree and direction of the change cannot be accurately ascertained unless we are familiar with the dimension of the existing technologies of our clientele. Second, it is generally believed that farmers experiences are time tested. Their technologies have stood the test of time in their respective ecological relationship, regardless, whether these were released by any institute or not. Third, utility of such studies lies in their potentiality to contribute problems for the much talked about field oriented research, one way to select field-oriented research could be to thoroughly diagnose the farmers' practice and spot out strong and weak points therein and explain their way. While we have research wing to know the lag in adoption of research, there is no standard method in operation to distil the problems from field soliciting research. Now a day's so many scientists and engineers who have graduated from India's premier institutions are advocating traditional scientific beliefs and methods. Several small, informal networks (Honey bee run by Anil Gupta, IIM, Ahmedabad) gave such innovators mutual support and means to learn from each other. There were too many people similarly advocating and working towards a return to the traditional wisdom.

Conclusions

Identification, documentation and validation of IKS of mountain regions are of paramount importance. An inventory of documented practices will enrich IK system by making the people aware about the indigenous technologies which are near to extinction and also provide opportunity to the scientists to appreciate and reorient their research programmes. Village Technology Centers should be established in the villages of mountain areas. These centers will provide technological support to village people at one hand and scientists and development workers at other hand. Training programmes on IKS should be launched for creating awareness and sensitizing rural people, rural youth and rural women towards IKS. Ultimately such indigenous knowledge systems will facilitate blending of modern scientific knowledge with traditional knowledge for sustainable development of mountain eco-systems.

Acknowledgement

Help rendered by Ms. Neelam Yadava, M.Sc. Scholar (Development Communication) of G.B. Pant University of Agriculture and Technology, Pantnagar in final computer setting of the manuscript is acknowledged.

References

Berkes, F. and C. Folke. 1994. Linking social and ecological systems for resilience and sustainability. Paper presented at the workshop on Property Rights and the Performance of Natural Resource Systems. Stockholms: The Beijer International Institute of Ecological Economics, the Royal Swedish Academy of Science.

Gautam, P.L. 2004. Livestock and mountain development–the way forward. In Vir Singh and P.L. Gautam (Edited) *Livestock Production Systems for Sustainable Food Security and Livelihoods in Mountain Areas*. GBPUA&T, Pantnagar.

Hecht, S.B. and Pscy, D.A. 1989. Preliminary results of soil management techniques of the Kayap Indians", *Advances in Economics Botany* (7): 174-188.

IUCN. 1997. *Indigenous Peoples and Sustainability: Cases and Actions*. Utrecht International Books.

Maundu, P. 1995. *Methodology for Collecting and Sharing Indigenous Knowledge: A Case Study, Indigenous Knowledge and Development Monitor*, Vol.3 (2).

Melkania, N.P. 2002. Forage and livestock in mountain farming systems in India: issues and perspectives. *Souvenir and Abstracts of IYM Symposium on Livestock Production Systems for Sustainable Food Security and Livelihoods in Mountain Areas*. Organized at GBPUA&T., Pantnagar during 30-31 December, 2002.

Morin-Labatut, G.M. and S. Akhtar. 1992. Traditional environmental knowledge: a resource to manage and share", *Development Journal of the Society of International Development* 4: 24-29.

Salas, M.A. 1994. The technicians only believe in science and cannot read the sky, pp. 57-69 in Scoones and J. Thompson (Eds) *Beyond Farmer First: Rural People's Knowledge, Agricultural Research and Extension Practice.* London: IT Publications.

UNEP. 1992. *Inter-Commission Tasks Force: Convention on Biological Diversity.* Geneva Executive Centre, Chatelaine Switzerland.

Van Vlaenderen, H. 1999. Local knowledge: what is it, why and how do we capture it? In *Gender, Biodiversity and Local Knowledge Systems (Links) to Strengthen Agricultural and Rural Development Paper* (GCP/RAF/338/NOR), Morogoro.

Wang, C. 1988. Indigenous Communication System in Research and Development. *Journal of Extension Systems* 4 (2): 75–86.

Warren, D.M. 1991. *Using Indigenous Knowledge in Agricultural Development.* World Bank Discussion Paper No.12, Washington D.C., The World Bank.

Chapter 15

Traditional Management of Range Resources:

Mountain Farmers' Ecological Tactics in the HKH Region

☆ *Vir Singh and R.D. Gaur*

Rangelands are the uncultivated areas or abandoned cultivated areas and occasionally fallow lands that harbour natural or seeded/ planted vegetation of herbaceous and woody species and that serve as habitats of a variety of wild animals, support domesticated animals, provide unique products and ecosystem services especially vital for the livelihoods of livestock-dependent communities (Singh 2007). Rangelands constitute the largest geographical area in the mountains of the Hindu Kush-Himalayan (HKH) Region. These ecosystems play key role not only in imparting the ecological integrity to the mountain agro-ecosystems but are also of critical value in the life of people. Mountain farmers have adopted and practice ecological tactics of range resource management (conservation and utilization). An attempt has been made to elaborate these management tactics that emanate from the treasure of ecological wisdom of range- and livestock-dependent farmers in the mountains throughout the HKH Region.

Animal husbandry is older than crop cultivation. Human civilization, in fact, entered into a new phase of its development and evolved to the present state thanks to the adoption and practices of animal husbandry. Range management appears to be as old as animal husbandry. And so is the knowledge relating to range management. Range management also appears to be the oldest art of natural resource management. Since the dawn of civilization, natural resource management systems have been the local community-based systems. Indigenous perceptions and practices regarding management of natural resources primarily concerned with protection and utilisation of biologically diverse resources. Human-managed natural resource systems endowed with nature's biodiversity virtually are as diverse as the human-inhabited ecosystems on Earth.

The Himalayan region has unique functions and roles, and range management by local farmers and transhumant societies is the largest intervention into it. Throughout the history the interaction between mountain people and the natural systems had helped communities to maintain the richness of species, communities and genetic materials in both the productive systems and wild lands of the mountain environments (Shengji, 1994).

There have been three fundamental means to manage rangelands in the Himalayan mountains, particularly in its north-western region: i) burning of the forests; ii) cutting down and/or thinning of the forests; and iii) abandonment of cultivated lands.

Amongst these, the former has been the method that could be easily handled by people. Slash-and-burn farming, also called the swidden farming, shifting farming or *jhoom* cultivation in the north-eastern hills, is different from that practiced in the north-western Himalayan mountains. While in the former case the cleared forest area is used for cultivating food and/or cash crops and place of agricultural activities is changed after a cycle of few years, in the latter, the forest area is managed for pastoral activities and it is a part of the settled agriculture.

Cutting down or thinning of a forest area is an old art of silvi-pasture management. It involves an appropriate mix of woody species and ground flora of herbaceous plants dominated by grasses. Such a management of specific vegetation involving economic species to provide specific products (foods, fodder, fuel wood, fibre, timber, medicines, etc.) and specific functions (soil and water conservation, micro-climate maintenance, synergy, etc.) in the rangeland ecosystems is a wonderful articulation of local farmers' traditional knowledge.

Burning and cutting down and/or thinning of forests allows the herbaceous species (Poaceae and non-Poaceae families) to thrive on ground exposed to more solar radiation. These species are of special socio-economic importance. In a rangeland, the first preference is given to the fodder species. The ground flora is virtually used as fodder for livestock. However, woody species occurring in a rangeland ecosystem, through their leaf biomass, also contribute significant amount of fodder to farmers for their livestock. All woody species are often used for extracting fuel wood while many others may be of multiple economic uses simultaneously providing other useful products such as fruits, edible seeds, dyes, medicine, fibre, bedding material, minor timber, agricultural implements and tools, etc.). Some species may be of little direct economic importance but are maintained to render useful ecosystem services vital for the maintenance of agricultural and livelihood systems in mountain areas.

Abandoned agriculture on a piece of land also results in the development of a rangeland ecosystem. This land is often predominated by grasses and other herbaceous plants interspersed with shrubs and trees, which were originally kept as part of agro-forestry system. The trees and shrubs are broadleaved extensively used for top lopping.

There is yet another practice of farmers in which cultivated land is used as grazing ground during certain period of the year. This is the fallow practice. Terraced fields used for cultivation are left fallow for a crop season in a year. During this season, livestock can graze on the fallow fields. Grasses of field bunds, green plants, or weeds, growing in the absence of ploughing, leftovers of the previous food crop (leaves, stems and often roots) are grazed by the animals. The private cropland left fallow is also used as CPR during the fallow period. Fellow practice leads to increase in grazing area temporarily. This is the practice adopted by farmers for centuries to give periodic rest to their fields for maintaining fertility of the soils. The soil fertility through fallowing practice is managed by i) preventing draining out of nutrients via food crops, ii) allowing microbial degradation of crop residues (mostly roots), iii) *in-situ* manuring by the grazing animals, and iv) green manuring upon mixing up green plants/weeds through ploughing at the end of fallow period.

Indigenous knowledge (IK) is the knowledge of the people living in a certain area, generated by their own and their ancestors' experience and including knowledge originating from elsewhere but has been practically accepted by the local people. It is also referred to as local knowledge (Reijntjes *et al.* 1992).

Rural people of the HKH Region are rich in the knowledge about the native plants, most of which are occurring in uncultivated (forest/rangeland) areas. Rare examples relating to Himalayan ethnobotany were highlighted by Gaur (1999, 2004) and Gaur *et al.* (1984, 1988, 1992) indicating the prevalence of indigenous knowledge in the Region.

Site-Specific Management

The natural resources, techniques and strategies from which farmers can choose in creating, maintaining and developing their farm systems involving rangelands are largely determined by the specific characteristics of the ecosystems. Local farmers have normally investigated local options to the greatest possible extent and exploit them well. In this way, finely tuned farming systems have developed which have survived for generations. The local farming community relies on physical and biological diversity. Advantage is taken of complementarity between crops, animals and pastures. Each farming community has developed different natural resource/rangeland management techniques that suit the specific local conditions (Reijntjes *et al.*, 1992).

Local farmers used to manage a blend of diversity of trees, shrubs, and herbaceous plants by conserving, utilizing, planting, and protecting in accordance with the site specificities. They are well acquainted with a particular site and the type of vegetation that can grow over there. Altitude, slope-orientation, and other topographic or physiographic factors–including edaphic factors and microclimate–are the main basis of the livestock-dependent local farming communities to manage their rangeland resources.

Integrating Range and Livestock

Available rangeland vegetation is converted into milk, meat, wool, draught power, etc., by means of rearing animals and using range vegetation as source of their feed. This helps the farmers increase subsistence security by diversifying the food generating and other economic activities. In addition, this integration also helps transfer energy nutrients between animals and cultivated crops via manure and via use of draught power. Livestock can enhance cropland productivity by intensifying energy and nutrient flows.

Rangeland-livestock integration is one of the key strategies of the livelihood systems evolved by mountain farmers. A rangeland ecosystem stocked with enormous biodiversity of trees, shrubs, herbaceous plants, wildlife, and micro forms of life is a rich store of energy and nutrients and is ecologically more stable than a cropland. Biodiversity of rangeland ecosystems has its bearing on the biodiversity of food crops in cropland areas thanks to the range-livestock-farming integration. By means of the diversification of food production activities ensured through this integration, a farming community enhances the level of its food security.

Livestock Management

Mountain farmers rear several species of livestock to derive full advantages of the available natural resources in their environments. By maintaining diverse species–*e.g.*, cattle, buffaloes, goats, sheep, yaks, horses, mules, ponies, rabbits, poultry, etc.–farmers tend to exploit a wider range of feed resources. This might not have been possible with one or two species. Cattle prefer grasses and herbs and so do sheep. Sheep, however, would resort to browsing during the scarcity of the preferred fodder. Goats are browsing animals. They eat the leaves of shrubs and trees. Buffaloes generally prefer stall feeding, but transhumant communities, like Gujjars, manage large number of buffaloes under grazing over vast tracts of rangelands from the foothills to alpine zone of the Himalaya. Yaks take full advantage of the alpine meadows adjacent to perpetual glaciers. Poultry birds can thrive on foraging around homesteads and on kitchen waste.

Transhumant pastoralists select the species and breeds of animals that are capable of walking long distances and covering treacherous and rugged terrain. The animal breeds farmers rear are suitable and well-adapted to mountain environments, can subsist on local feed resources, have high feed conversion efficiency and resistance against diseases.

Ethnovet medicine

Conventional animal health management is exorbitantly expensive and unaffordable by small and marginal farmers who largely depend on their livestock resources for their livelihoods. The modern health cover essentially requires external veterinarians who are not well acquainted with the problems of some of the inaccessibility- and poverty-ridden areas, such as the Himalayan mountains. Ethno-veterinary practices in the Himalayan mountains areas are not only common and form a dominant system of livestock health management, but also offer a cornucopia of benefits to the poor livestock-dependent communities as well as to the entire livestock sector. Remote, inaccessibility-ridden poor areas inhabited largely by tribal and pastoral communities are especially rich in the ethnovet-related knowledge. In the garb of modern veterinary advances, such traditional knowledge systems appear to have lost ground and hence need to be recognized, re-examined, revived and promoted. Gaur *et al.* (1992), Tripathi and Mandape (2002), Joshi and Mandape (2004), Joshi and Singh (2004), Mandape *et al.* (2004), Joshi *et al.* (2005) and Joshi and Singh (2006) have documented some ethnovet practices prevailing in the Indian rural areas of Uttarakhand Himalaya. Gaur (1999) has given panorama of the plant diversity present in diverse ecosystems in the Garhwal district of Uttarakhand Himalaya, along with ethnobotanical notes of each species.

There are a number of medicinal plants occurring in diverse ecological niches which are of great ethnovet value and used for treatments of various ailments by local farmers. In mountain areas, *Aegle marmelos, Aloe vera* and *Trigonella foenumgraecum* plants are used for the treatment of burns. *Annona squamosa, Azadirachta indica, Chenopodium album, Eclipta alba* and *Lens culinaris* are vital for the quick recovery from fresh and cut wound. *Caryopteris odorata, Girardinia diversifolia, Grewia optiva, Murraya paniculata* and *Pinus roxsburghii* are used for the treatment of bone fractures. *Asparagus officinalis* plant is utilized in increasing milk production in lactating mothers and is also used in health tonics. *Ocimum sanctum, Melia azadirachta, Tagetus erectus, Hibiscus rosa chinensis, Phyllanthus niruri, Crinum asiaticum, Vetiveria zizaniodes, Cannabis sativa, Bacopa monnieri,* etc. are other plants of ethnovet value extensively used by farmers in the mountains (Joshi *et al.*, 2005).

In the Terai and plain areas, *Triginella foenumgraecum, Trachyspermum ammi, Carcica papaya, Zinziber officinale, Allium sativum* and *Saccharum officinarum* are used for the treatment of respiratory disorders in livestock. *Euphorbia hirta, Pistia stratoites, Syzigum jambolanum, Cyamopsis psoralioides, Cuscuta reflexa* and *Cuminum cymium* are vital for curing digestive disorders in livestock. *Acacia arabica, Brassica nigra, Azadirachta indica* and *Allium sativum* are used in skin disorders. *Allium cepa* and *Tinospora cordifolia* are used for curing fever while *Ficus bengalensis* and *Cocus nucifera* are used for curing swelling neck. *Aegle marmelos* and *Ricinus communis* are used for evacuation of dead calf (Joshi *et al.*, 2005).

Tribal belts of the Greater Himalayas are especially rich in the ethno-vet practices. The environments in this area are especially suitable for an exceptionally rich variety of medicinal plants vital for the treatment and control of common and rare livestock diseases. Use of the rare herbs is made in case of pregnancy, at the time of parturition, for sustaining milk yields, improving quality of livestock products as well as for the management of specific health problems. Some herbs are common for human and livestock health management (Joshi *et al.*, 2005; Joshi and Singh, 2006).

Knowledge regarding specific medicinal plants and their usage is on the tips of the livestock farmers. Farming communities inhabiting Terai and plains of the Indo-Gangetic valley have lost much of the knowledge relating to the medicinal plants and their usage in health management. Some old people, however, would readily identify the plants and their uses in specific diseases/ problems.

New generation is more dependent on modern vet medicines. Ethno-veterinary practices go hand in hand with traditional farming. Transformation of agriculture leads to the erosion of these practices. The other dominant factor is the degree of inaccessibility. Higher the degree of inaccessibility, as in the mountain areas, higher would be the dependence on ethno-medicines (Joshi *et al.*, 2005).

Ethno-vet practices of the farming class have been evolved over millennia through trials and errors. They should be regarded as futuristic rather than condemned as a remnant of the past. Recognition, documentation, demystification, augmentation, promotion and articulation of this traditional knowledge system would essentially contribute to the amelioration of livestock-based production systems and economic returns to the farmers.

Grazing Management

Ecologists today often make advocacy for deferred-cum-rotational grazing. The pastoralists in the Himalayan areas, however, have been adopting this practice for decades, probably for centuries. They do not graze their animals on a single patch in a limited area, for they are well aware that frequent defoliation eventually ends up in the cessation of plant regeneration. They keep shifting grazing area and closing the earlier grazed area for giving enough time for plant regeneration, and so on. The transhumant communities are especially conscious about changing grazing grounds according to seasons and thus about the maintenance of rangelands' health.

Browsing animals, such as goats, are useful in reducing encroachment of unwanted bushes into pasture. The selective grazing habit of animals can be deliberately used to control 'weeds'.

Fields left fallow are used as CPRs, especially for grazing animals. While grazing, animals void dung and urine and fertilise the soil. This practice is called kraaling. In Nepal and parts of the Indian Himalaya, farmers have to pay the owners of goats and sheep against the kraaling. In-situ manuring, *i.e.* the manuring of field by tethering the animals directly in the fields is an important strategy developed by the mountain farmers over ages. During dry season, cattle (and small ruminants) graze harvested fields of finger millet left fallow for fertility regeneration and manure them directly. The draught animals also directly manure the fields, in which they perform agricultural operations by voiding their dung and urine. Results of an experiment by Powell and Ikpe (1992) showed that millet yields in kraaled areas (which received both manure plus urine) were over twice as high as yields where only manure from stall-kept animals was spread by hand. Residual benefits of urine were also observed. Yields in plots where cattle and sheep were kraaled the previous year were about 1.5 times greater than where only manure had been hand-spread (Powell and Ikpe, 1992).

Transhumant pastoralists and nomads follow their own rules. They would sometime migrate even to such distant areas which are owned and governed by different states or nations. Nomads of Jammu and Kashmir would visit the rangelands in the state of Himachal Pradesh in India and those in Himachal Pradesh would be found grazing their animals in Uttarakhand mountains. Nomads of Afghanistan utilize range resources of Pakistan and vice-versa. Nomads visiting other's pastures are often accepted and welcomed.

Grass Harvesting

Vegetative propagation and sprouting of seeds are the two means of plant regeneration in the ecosystems. Regeneration by seed shedding is taken advantage of so that root reserves of energy are not exhausted by frequent cuttings. The grass is harvested after shedding of seed in the months of October-November. This is the natural way of allowing ecological regeneration of rangelands of which the farmers are well acquainted.

Grass Distribution

Managed in protected plots in the village rangelands, the grass is distributed to all families on the basis of some egalitarian principles, and social sanctions work in this regard. Harvesting is to be done by the family members, invariably by women, who have to carry the load to home. A family would bring equal numbers of bundles in a day until all the grass is harvested. Sometime all families would be allotted equal number of days for harvest and bring home as much grass as one can during those days. Sometimes this arrangement is done that a woman of a family would bring as many head loads of grass as she can during a day harvest. Sometimes numbers of animals with a family would be the criterion of distribution. Grass distribution pattern thus would depend on the mutual decision of the people of an area or local governance body, *e.g.* a *Gramsabha* in Indian Himalayan Region. Nomads often do not require collection of grass from a certain area. They inhabit areas (*e.g.*, forests) of the foothills during winter season. Such areas provide fodder even during the scarcity period.

Livestock Feeding

Farmers have their own strategies of livestock feeding. They have deep knowledge about the nutritive value of the feeds and fodders they offer to their animals. *Bhimal* (*Grewia optiva*) and *kharik* (*Celtis australis*) are regarded to be the best tree-based green fodders. Their knowledge about these two tree fodders is in conformity with the published reports on the nutritive values of tree fodders of mountain areas. These two trees provide leaves with high crude protein value free from incriminating factor tannin and with high digestibility value (Singh and Naik, 1987; Singh *et al.* 1995; Singh and Bohra, 2005). They would also prefer other tree species like *Bauhinia, Ficus, Ougeinia, Vitis,* etc. which are also known to offer highly nutritive fodder.

Farmers select appropriate fodders in different seasons and try to make balance of nutrients in livestock feeding. They have all knowledge relating to the lopping cycle or fodder calendar. They respond to lean period (period of fodder scarcity) by way of incorporating leaves of evergreen trees and/or field weeds and by nutrients supplement through concentrate feeding.

Farmers prepare their own concentrate, the feed rich in total digestible nutrients and energy, incorporating home-produced food grains. The ingredients used are black soybeans (10 per cent), soybeans (15 per cent), barnyard millet (10 per cent), gram (5 per cent), wheat (45 per cent), barley (10 per cent) and sugar (5 per cent). Locally called *chaata* in the Kumaun Himalaya, this concentrate is supplemented in the diet of pregnant and milking animals and of working bullocks.

Farmers provide dietary allowance to pregnant cows and buffaloes and to milking animals. This is provided through concentrate supplement. Dry animals (*i.e.,* animals not in milk) are seldom provided concentrate. Bullocks during idle days are also not fed concentrate. They, however, are given nutrients rich diet during the days when they are put to work. Goats and sheep are kept exclusively on grazing.

Farmers offer specific feeds and fodders to increase quantity and quality of milk. The straws of *Echinochloa, Oryza, Hordeum, Eleusine, Triticum* and *Pennisetum* are given to livestock, which is an age old practice. There are straws of a number of cultivated plants which do not make any impact on milk production. Besides the straws of cereals and legumes, there are several plants used for fodder. These plants directly affect the quantity as well as quality of milk. Local inhabitants are well acquainted about it. Several wild plants increase the nutritional value of the milk and lopped severely for the purpose. On the other hand, there are some plant species which often decrease the quality of milk. Such types of species are used as forage during scarcity or famine period.

Gaur *et al.* (1983) and Negi *et al.* (1985) have presented a list of the plant species in Garhwal Himalaya that increase or decrease the quantity and quality of milk. Some of these plants are being named below.

Plant species that increase milk quantity and quality and identified by the farmers are: *Acacia farnesiana* (L.) Willd., *Bauhinia purpurea* L., *B. vahlii* Wight and Arn., *B. variegata* L., *Bothriochloa glabra* (Roxb.) A. Camus, *Carpinus viminea* Lindley, *Cassia tora* L., *Caesalpinia decapetala* (Roth.) Alton, *Castanopsis indica* Roxb., *C. tribuloides* (J.E. Smith) A. DC., *Celtis australis* L., *Chenopodium album* L., *C. murale* L., *Chrysopogon gryllus* (L.) Trinius, *Cissampelos pareira* L., *Clematis barbellata* Edgew, *C. grata* Wallich, *Crotolaria medicaginea* Lam., *Cyananthus lobatus* Wallich ex Benth., *Cynodon dactylon* (L.) Persoon, *Danthonia jacquemonti* Bor., *Debregeasia salicifolia* (D. Don) Rendle, *Dichanthium annulatum* (Forsk.) Stapf, *D. aristatum* (Foir.) C.E. Hubb, *Digitaria cruciata* (Nees ex Steudel) A. Camus, *D. setigera* Roth apud Roemer et Schultes, *Ehretia laevis* Roxb., *Euphorbia hirta* L., *E. pilosa* L., *Festuca kashmeriana* Stapf, *Ficus glomerata* Roxb., *F. neriifolia* Smith var. *nemoralis* Wallich ex Miq., *F. palmata* Forsk., *F. roxburghii* Wallich ex Miq., etc.

Plant species which contribute to decrease quantity and quality of milk are: *Andropogon echoides* L., *Artemisia nilagirica* var. *septentrionalis* (C.B. Clarke.) Pampanini, *A. parviflora* Roxb. ex D. Don, *Codonopsis ovata* Benth., *Coriaria nepalensis* Wallich, *Erianthus munja* Roxb., *Erythrina variegata* L., *Fagopyrum cymosum* Treviranus, *Hedera nepalensis* K. Koch, *Hemarthria compressa* (L.f.) R. Br., *Hippophae salicifolia* D. Don, *Hypericum elodeoides* Choisy, *H. hookerianum* Wight and Arn., *H. nepaulense* Choisy, *H. oblongifolium* Choisy, *Impatiens scabrida* DC., *Myrsine semiserrata* Wallich, *Phacelurus speciosus* (Steudel) C.E. Hubbard, *Urtica dioica* L., etc.

Fodder Preservation

Preservation of high quality forages by drying is known as hay making. Hay making is the simplest way of fodder preservation and mountain farmers have been using this technique for centuries. In India sunlight is available plentifully for most of the days during the year. Farmers dry the green forage in sunlight in the open. This is the most economic method of preservation.

Hay making involves reduction of moisture content through sun-drying or artificial drying to 10-20 per cent which inhibits enzyme activity in the plants. Mountain farmers observe this scientific principle of hay making.

The harvested grass is preserved to be fed to livestock during winter season. People are well aware of the natural preservation methods, a technology which is centuries old. The grass is spread loose often on rooftops or on the broad fence around homestead and sun-dried for a couple of days. Moisture percentage is kept about 10-15 percent. Also recommended by the animal scientists, this moisture level curbs microbial growth in the plants that could take place at appropriate moisture level, and thus helps preserve the fodder for longer duration.

Nutritive value of fodder is also taken care of in a scientific manner by the farmers when applying natural preservation technology. Leaves of the plants, which are more nutritive than the stem, are not allowed to shatter. Sun-drying also helps increase vitamin D content in the fodder.

The preserved fodder is stacked on the branches of trees or tied around a wooden pole near homestead. The stacking method is also a form of an appreciable art. The stacked fodder, locally called *luta* in Kumaun and *Purolda* in Garhwal, is given the shape of a temple, broad base ending into a conical tip. This shape, 1.0 to 1.5m straight height, allows the rain water flow down the preserved material quickly with minimum seepage, thus helping prevent the fodder from getting spoiled by increasing moisture content. In mountain areas there is no enough space that could be spared for storage of fodder. The traditional fodder stacking/storage technique of farmers is an appropriate response.

Traditional knowledge relating to rangeland management in the HKH Region is summarized in Table 15.1. This management is ecologically sound. Various approaches involving in range resource management are of crucial ecological and socio-economic implications (Figure 15.1). Traditional management (conservation and utilization) as being practiced in the HKH Region for millennia should not be overlooked in institutional development intervention.

Table 15.1: Indigenous Technology of Range Management in the HKH Region

Management Relating to:	*Indigenous Technological Applications*	*Logic*
Site Specificity/ Niche Ranges	☆ Diversity at ecosystem level; ☆ Niche-specific vegetation; ☆ A mix of trees, shrubs and herbaceous plants.	☆ Diversity of ecological niches according to altitude, slope direction, amount of precipitation, micro-climate, etc. is naturally available in the mountains (*e.g.,* Dune ranges, forest grazing lands, mid-altitude ranges, alpine pastures, trans-Himalayan ranges, etc.); ☆ Niche-specific vegetation (e.g., subtropical forages, temperate vegetation, alpine meadows, or sal type, oak type, alpine type, etc.) creates diversity amongst and within species that is used in range management and subsequently in nourishing livestock-dominated local livelihoods; ☆ A mix of vegetation types makes the rangeland ecosystems, more heterogeneous, ecologically more stable, more resilient and more productive.
Floral Composition	☆ Periodic burning;	☆ Allows the herbaceous species (gramineae and non-gramineae families) to thrive on ground exposed to more solar radiation;
	☆ Cutting down and/ or thinning of forests.	☆ Dominance of the vegetation of greater socio-economic importance; ☆ More amount of fodder flow from the ecosystems for raising livestock; ☆ Control of undesirable weeds through selective grazing;

Contd...

Table 15.1–Contd...

Management Relating to:	*Indigenous Technological Applications*	*Logic*
		☆ Maintenance of species of intangible uses contributing to vital ecosystem services; ☆ Using plant-animal interactions for soil fertility management in croplands and for food production.
Cultivated Lands	☆ Abandonment of agriculture on poor lands; Practice of periodic fallowing.	☆ Appropriate land use; ☆ More efficient use of cultivated land; ☆ Use of PPRs as CPRs that enhances socio-cultural cohesion within a community; ☆ Improvement in the fertility status of soils of the fallow lands through preventing draining out of nutrients via food crops, allowing microbial degradation of crop residues (mostly roots), *in-situ* manuring by the grazing animals, and green manuring upon mixing up green plants/ weeds through ploughing at the end of fallow period.
Livestock Keeping	☆ Native breeds of many species;	☆ Diverse species to take advantage of diverse agro-ecological zones;
	☆ Transhumance.	☆ Livestock breeds with high feed conversion efficiency, adaptability and resistance against diseases, and capable of traveling long distances; ☆ Transfer of energy and nutrients from more stable range ecosystems to more fragile croplands via livestock.

Contd...

Table 15.1–Contd...

Management Relating to:	*Indigenous Technological Applications*	*Logic*
Ethno-vet Medicine	☆ Use of various plants occurring in mountains' diverse ecological niches as medicine for livestock health management.	☆ Ethno-medicines offer a cornucopia of benefits to the poor livestock-dependent communities as well as to the entire livestock sector; ☆ Suitable for remote, inaccessibility-ridden poor areas inhabited largely by tribal and pastoral communities; ☆ Cost-effective and safe approach to livestock health management; ☆ No side-effects of medicine; ☆ Availability in the vicinity of households.
Grazing	☆ Rotational-cum deferred grazing of livestock; ☆ Fencing; ☆ Enclosure of areas/plots for harvest of grass at appropriate time for stall-feeding of livestock.	☆ Avoidance of frequent defoliation and trampling of plants for quick regeneration of palatable plants; ☆ Plots left for harvesting grass for feeding during winter season; ☆ More appropriate use of uncultivated lands.
Fodder Distribution	☆ Distribution of grass before harvest to all families on equal basis following social sanctions; ☆ Harvest by individual families.	☆ Egalitarian distribution indicative of benefit sharing and social justice maintained through social sanctions; ☆ Strengthening of local governance; ☆ Appropriate participatory management of CPRs.

Contd...

Table 15.1–Contd...

Management Relating to:	*Indigenous Technological Applications*	*Logic*
Lopping and Harvesting	☆ Only about one-third of the tree canopy is lopped ☆ Annual cutting of grass after seed set and seed shed	☆ Ecological regeneration of plants through vegetative propagation and germination of seeds shed in the grass plots; ☆ Regeneration by seed shedding is taken advantage of so that root reserves of energy are not exhausted by frequent cuttings.
Livestock Feeding	☆ Choosing tannin-free species with high crude protein and digestibility value; ☆ Response to seasonality and lean period; ☆ Preparation of concentrate incorporating locally produced food grains; ☆ Giving dietary allowance according to the physiological states of livestock; ☆ Preferring fodders that increase quantity and quality of milk.	☆ Tannins are incriminating factors that decrease digestibility of crude protein; ☆ Livestock do not often face paucity of nutrients in their diet during scarcity period; ☆ The concentrate has high energy and protein value and is rich in overall digestible nutrients having bearing on the health, maintenance and productivity of livestock; ☆ Dietary allowance takes care of the wear and tear of animal tissues and caters to the need of extra nutrient supplies during pregnancy, milk production and muscular work by the animals; ☆ Increasing quantity and quality of milk through selective/ balanced feeding has implications on family's health and economy.

Contd...

Table 15.1–Contd...

Management Relating to:	*Indigenous Technological Applications*	*Logic*
Fodder Preservation	✰ Sun-drying of grass left loose for appropriate period to reduce moisture level to 10-15 per cent; ✰ Stacking of unchopped fodder in temple shape on tree branches or wooden poles.	✰ Maintenance of nutritive value by keeping leaves intact with the plant; ✰ Preventing fodder from spoilage by microbial growth to help preserve the fodder for longer duration; ✰ Increase of vitamin D content in fodder; ✰ Saving from the rain water by allowing minimum seepage.

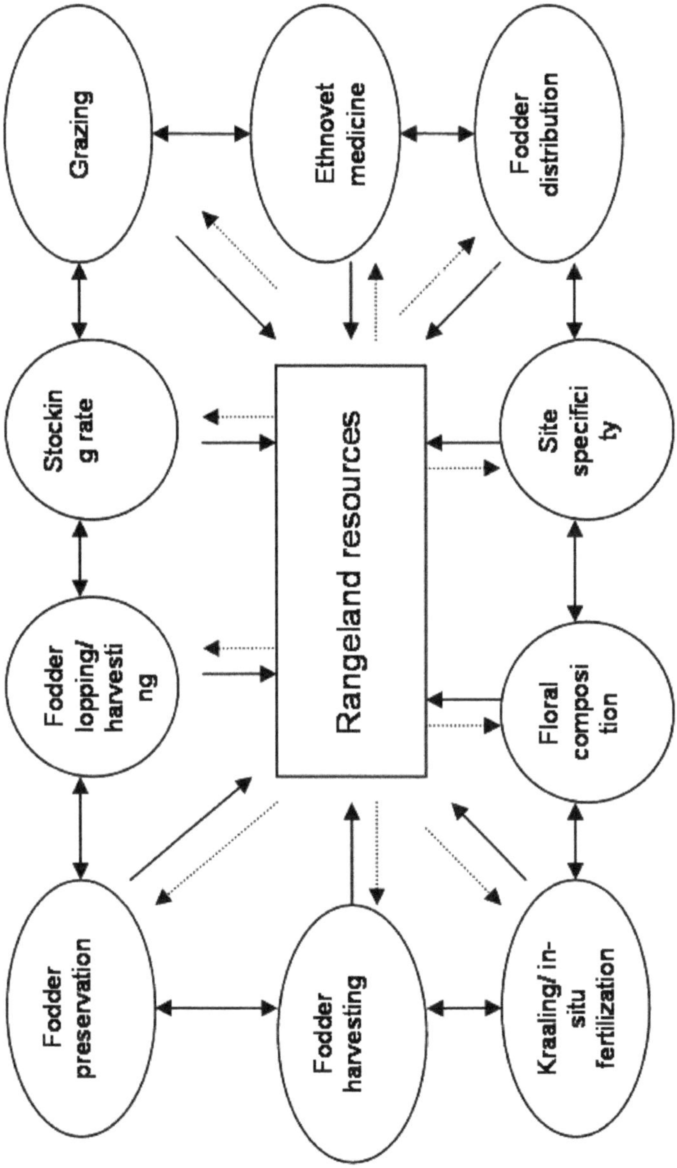

Figure 15.1: Implications of Range Resource Management Based on Ecological Tactics

This could be instrumental towards evolving sound strategies of ecodevelopment in the fragile ecosystems.

References

Gaur, R.D. 1999. *Flora of the District Garhwal, North West Himalaya (with Ethnobotanical Notes)*. Transmedia, Srinagar, Garhwal.

Gaur, R.D. 2004. Vegetational diversity of Uttarakhand Himalaya and scope of regional development. In *Resource Appraisal, Technology Applications and Environmental Challenges in Central Himalaya*. HNB Garhwal University, Srinagar.

Gaur, R.D., Bhatt, K.C. and Tiwari, J.K. 1992. An ethnobotanical study of Uttar Pradesh Himalaya in relation to veterinary medicines. *J. Indian Bot. Soc.*, 72: 139-144.

Gaur, R.D., Bisht, M.K. and Bhatt, K.C. 1988. A note on medicinal and other uses of some animal in Garhwal Himalaya: An ethnozoological perspective. *Man in India*, 68: 292-297.

Gaur, R.D., Semwal, J.K. and Negi, K.S. 1983. Traditional fodder yielding plants of Garhwal Himalaya and their impact on quality and quantity of milk. *JOSHRAD*, 6: 89-93.

Gaur, R.D., Semwal, J.K. and Tiwari, J.K. 1984. A survey of high altitude medicinal plants of Garhwal Himalaya. *Bull. Med. Ethnobot. Res.*, 4(3-4): 102-116.

Joshi, M. and Singh, V. 2004. Ethnobotanical valuation of some plants in Himalayan Environments. In: *Environmental and Pollution Awareness in Hilly Areas*, Jyothi Prasad and K.N. Shukla (eds.), G.B. pant University of Agriculture and Technology, Pantnagar.

Joshi, M. and Singh, V. 2006. Ethnobotanical importance of some endemic plants in Central Himalayas. In Rawat, M.S.S. (ed.) *Resource Appraisal, Technology Application and Environmental Challenges in Central Himalaya*, pp. 192-199. Department of Geography, HNB Garhwal University, Srinagar (Garhwal).

Joshi, M., Bohra, B., Mandape, T., Joshi, S. and Singh, V. 2005. Ethno-veterinary practices in rural areas in India. In Rowlinson, P., Wachirapakorn, C., Pakdee, P. and Wanapat, M. (eds.) *Integrating Livestock-Crop Systems to Meet the Challenges of Globalisation*. British Society of Animal Science, U.K.

Joshi, S. and Mandape, T. 2004. Hill Plants to the Rescue of Human

Health: Evidence of Traditional Knowledge from Tehri Garhwal. In: Environmental and Pollution Awareness in Hilly Areas, Jyothi Prasad and K.N. Shukla (eds.), G.B. pant University of Agriculture and Technology, Pantnagar, India, 70-72.

Mandape, T., Joshi, S. and Singh, V. 2004. Use of ethno-phyto medicines in the hills. *Indian Farming*, 54(4): 22-23.

Negi, K.S., Tiwari, J.K. and Gaur, R.D. 1985. Economic importance of some common trees in Garhwal Himalaya: an ethnobotanical study. *Indian J Forestry*, 8: 276-279.

Powell, M. and Ikpe, F. 1992. Fertilizer factories: nutrient recycling through livestock. *ILEIA Newsletter*, 8: 13-14.

Reijntjes, C., Haverkort, B. and Water-Bayer, A. 1992. *Farming for the Future: An Introduction to Low-external-Input and Sustainable Agriculture.* Macmillan and ILEIA, Leusden, Netherlands.

Shengji, Pei. 1994. In Gurung, J.D. (ed.) *Indigenous Knowledge Systems and Biodiversity Management*. ICIMOD, Kathmandu.

Singh, V. 2007. *Studies on the Rangeland Management of Selected Areas in Kumaun Himalaya*, D. Phil. Thesis. HNB Garhwal University, Srinagar, India.

Singh, V. and Bohra, B. 2005. Livestock feed resources and feeding practices in hill farming systems: a review. *Indian Journal of Animal Sciences*, 75: 121-127.

Singh, V., Sharma, R.J. and Kumar, A. 1995. Perspectives on the utilization of forest fodder in the mountains. In Singh, V., Sharma, R.J., Kumar, A. (eds.) *Ecological Carnage in the Himalaya.* IBD, Dehradun.

Tripathi, H. and Mandape, M.K. 2002. Ethnoveterinary medical practices of Indian livestock owners in the northern plains of Uttar Pradesh. *Livestock International*, 6, 8-17.

Index

O

P

Q

R

S

T

www.ingramcontent.com/pod-product-compliance
Lightning Source LLC
LaVergne TN
LVHW020626300926
849161LV00004B/623